T0226116

Power Systems

More information about this series at http://www.springer.com/series/4622

Wen-Wei Chen · Jiann-Fuh Chen

Control Techniques
for Power Converters
with Integrated Circuit

 Springer

Wen-Wei Chen
Department of Electrical and Computer
  Engineering
Virginia Tech
Blacksburg
USA

Jiann-Fuh Chen
Department of Electrical Engineering
National Cheng Kung University
Tainan
Taiwan

ISSN 1612-1287 ISSN 1860-4676 (electronic)
Power Systems
ISBN 978-981-13-4983-6 ISBN 978-981-10-7004-4 (eBook)
https://doi.org/10.1007/978-981-10-7004-4

Printed on acid-free paper

This Springer imprint is published by Springer Nature
The registered company is Springer Nature Singapore Pte Ltd.
The registered company address is: 152 Beach Road, #21-01/04 Gateway East, Singapore 189721, Singapore

Contents

Chapter 1
Introduction

1.1 Power Converters with Integrated Circuits

Power converter technology is basic and important because it supports and delivers all types of electronic equipment and devices, such as consumer electronics, automotive, and telecommunication [1–11]. Switching and linear power converters are characterized by small size, light weight, low cost, high reliability, and high efficiency. Mainstream technologies and trend developments are used in power converters at present. Owing to the advancement of information and rise of a communication-oriented society, trends toward personalization and mobilization have become popular, and the demand for electronic devices that are small, light-weight, inexpensive, highly reliability, and highly efficient is increasing. Power converter technology is widely used to meet output loading requirements.

DC–DC power converters are utilized in portable electronic devices, such as cellular phones and laptop computers, which are primarily supplied with power by batteries or adapters. Such electronic devices often contain several sub-circuits to deliver auxiliary power, a central processing unit (CPU) voltage regulator, or a microprocessor, and the voltage level requirement of each differs from that supplied by the battery, adapter, or external power supply. In addition, battery voltage declines as its stored energy is drained. DC–DC power converters increase voltage from a partially reduced battery voltage and save space compared with using multiple batteries to accomplish the same task.

Figure 1.1 shows a brief circuit diagram of DC–DC power converters. DC–DC power converters can achieve power conversion from the input terminal to the output terminal. The input terminal of DC–DC power converters is the power source, and DC–DC power converters can regulate output voltage V_{OUT} to a specified level. Input capacitor C_{IN} is used to filter the input power source and is placed close to the V_{IN} pin of the integrated circuit (IC) on the evaluation board. This method can prevent the voltage drop of the printed circuit board (PCB) trace and ensures that the input voltage possesses good noise immunity. Output capacitor

© Springer Nature Singapore Pte Ltd. 2018
W.-W. Chen and J.-F. Chen, *Control Techniques for Power Converters with Integrated Circuit*, Power Systems,
https://doi.org/10.1007/978-981-10-7004-4_1

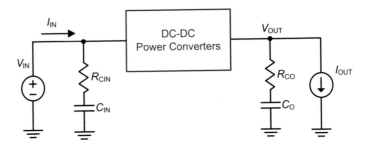

Fig. 1.1 A brief circuit diagram of DC–DC power converters

C_O is used to filter the output voltage and provide output capacitor C_O energy, thereby preventing output voltage V_{OUT} from dropping significantly at the load transient. Output capacitor C_O is placed close to the V_{OUT} or V_{FB} pin of IC on the evaluation board. This method can prevent the voltage drop of the PCB trace from affecting feedback voltage V_{FB} and ensures that feedback voltage V_{FB} possesses good noise immunity.

Power ICs are self-contained circuits with many separate components, such as transistors, diodes, resistors, and capacitors, fabricated into a single tiny chip of a semiconductor material [12, 13]. ICs are extremely small; they are thousands of times smaller than discrete circuits. ICs are also lightweight because of the miniaturized circuits and inexpensive because of the simultaneous production of hundreds of similar circuits on a small semiconductor wafer. An IC costs as much as an individual transistor because ICs are mass produced. ICs are highly reliable because of the absence of soldered joints and the need for only a few interconnections. ICs are widely used in DC–DC power converters because of their advantages.

DC–DC power converters with ICs are widely utilized in different industries, such as consumer electronics, automotive, telecommunication, networking, and medical. The various types of power ICs used at present include voltage regulators, battery management ICs, integrated application-specific standard product (ASSP) power ICs, and motor control ICs. Technological advancements and the increasing demand for battery-operated devices are the major factors that drive the market at present. The market is expected to grow because the demand for consumer electronics and automobiles is expected to increase in the future.

DC–DC power converters can be classified as step-down and step-up converters according to the difference between input and output voltages. A power converter whose input voltage V_{IN} is larger than output voltage V_{OUT} is called a step-down converter, whereas a power converter whose output voltage V_{OUT} is larger than input voltage V_{IN} is called a step-up converter. Several exceptions include high-efficiency light-emitting diode (LED) driver [14–19], which is a type of DC–DC power converter that regulates the output load current to drive LEDs. Another example is charge pumps, which are designed to generate an output voltage that is double or triple the input voltage.

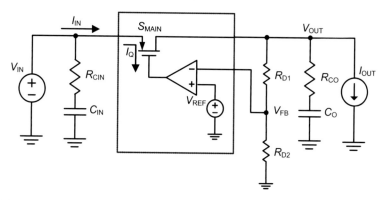

Fig. 1.2 Circuit diagram of LDO

The two types of DC–DC step-down converters are the switching buck converter and low-dropout linear regulator (LDO) [20–26]. Figure 1.2 shows a circuit diagram of LDO. LDO possesses a simple circuit structure because it usually consists of reference voltage V_{REF}, an amplifier, feedback resistors R_{D1} and R_{D2}, and power switch S_{MAIN}. LDO regulates DC voltage for the output voltage by reducing the input voltage across a power switch S_{MAIN}. The working principle of the LDO regulator involves sampling the output voltage through a resistive divider, which is fed to the inverting input of the error amplifier. The non-inverting input is tied to a reference voltage derived from an internal bandgap reference. The error amplifier always forces the voltages at its input to be equal. The voltage drop across the series power switch S_{MAIN} is controlled by the error amplifier's output to control the output voltage. The output voltage is expressed as Eq. (1.1).

$$V_{OUT} = \left(1 + \frac{R_{D1}}{R_{D2}}\right) \cdot V_{REF} \tag{1.1}$$

The advantage of LDO is that it requires only three parts: power switch S_{MAIN}, input capacitors C_{IN}, and output capacitors C_O. Several LDO IC products include resistors R_{D1} and R_{D2} inside their chip with a fixed output voltage. LDOs are usually cheaper and much less noisy than inductive switchers. Device input current I_{IN} is equal to the sum current between output load current I_{OUT} and quiescent current I_Q required by the LDO for its internal circuitry, as shown in Eq. (1.2). Quiescent current I_Q is the current drawn by the LDO to control its internal circuitry for proper operation. Series power switch S_{MAIN} and ambient temperature are the primary contributors to quiescent current. If the LDO is operated at a heavy load and output load current I_{OUT} is larger than quiescent current I_Q, quiescent current I_Q will not affect efficiency. However, if the LDO is operated at a light load and output load current I_{OUT} is smaller than quiescent current I_Q, quiescent current I_Q will affect efficiency and must be considered. The efficiency of the solution depends on the output-to-input voltage ratio in Eqs. (1.3)–(1.5). Given these features, the advantages of LDO linear regulators over other DC–DC regulators include the absence of switching noise, reduced IC die size, and design simplicity. Another advantage of LDO is its low cost. It does not require an inductor, so it saves cost.

$$I_Q = I_{IN} - I_{OUT} \tag{1.2}$$

$$P_{LOSS} = (V_{IN} - V_{OUT}) \cdot I_{OUT} + V_{IN} \cdot I_Q \tag{1.3}$$

$$P_{IN} = V_{IN} \cdot I_{IN} \tag{1.4}$$

$$\eta = \frac{P_{IN} - P_{LOSS}}{P_{IN}} \tag{1.5}$$

The main drawback of LDO is its low efficiency for high input-to-output voltage ratios or high-current applications. Most of the power is dissipated by power switch S_{MAIN} by Eq. (1.6). Power dissipation P_{PS} depends on the difference between input and output voltages and output load current I_{OUT}. Therefore, LDO is not an ideal solution when the difference between input and output voltages is large and when the output load current is high. Applications with large output load current require heat sinking, which increases the solution size. For these reasons, LDO is widely used for applications with low input-to-output voltage ratios and low current, such as auxiliary power delivery. Auxiliary power delivery is used to obtain a light load and high noise immunity with a high power supply rejection ratio (PSRR), as shown in Eq. (1.7). Although LDO obtains a low output voltage ripple without a switching voltage ripple, a good PSRR must be considered when selecting a suitable LDO.

Figure 1.3 shows the perturbation signal injection circuit at the input voltage to measure PSRR for LDO. AC perturbation signal V_{AC} can be injected by a transformer and the bandwidth of the transformer should cover a range between 100 Hz and 1 MHz to obtain a correct PSRR bode plot. To achieve impedance matching, the output terminal should be placed at 50 Ω parallel to the transformer. To obtain a correct PSRR bode plot, input capacitors C_{IN} should be removed so that the LDO

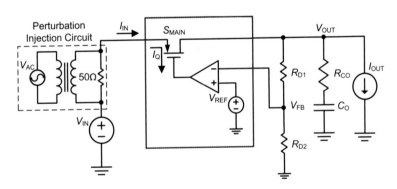

Fig. 1.3 Perturbation signal injection circuit at the input voltage to measure PSRR for LDO

can obtain its own PSRR without the capacitor's influence. Otherwise, AC perturbation signal V_{AC} may be filtered by input capacitors C_{IN}. The two output terminals of a network analyzer or spectrum analyzer should be connected to terminals V_{IN} and V_{OUT} to obtain a PSRR bode plot individually.

$$P_{PS} = (V_{IN} - V_{OUT}) \cdot I_{OUT} \tag{1.6}$$

$$PSRR = 20 \cdot \log \cdot \left(\left| \frac{V_{IN\_Ripple}}{V_{OUT\_Ripple}} \right| \right) \tag{1.7}$$

Most LDO IC products have a relatively high PSRR at a low frequency range of 100 Hz–1 kHz. Having a high PSRR over a wide band allows LDO to reject high-frequency noise, such as that arising from power switch S_{MAIN}. Similar to other specifications, PSRR fluctuates over frequency, temperature, current, and output voltage. Figure 1.4 shows the experimental results of PSRR for LDO at different output load currents (1 and 100 mA). The red-colored curve represents the measurement results for the PSRR bode plot at an output load current of 1 mA, and the blue-colored curve represents the measurement results for the PSRR bode plot at an output load current of 100 mA. The operation conditions are as follows: 3.3 V of input voltage V_{IN}, 2.5 V of output voltage V_{OUT}, 10 mV of AC perturbation signal V_{AC}, and 1 and 100 mA of output load current. At the output load current of 1 mA, LDO can obtain a PSRR of 60 dB, which is similar to that at the output load current of 100 mA.

At the frequency range of 100 Hz–450 kHz, the range of the PSRR results extends from bandgap filter roll-off frequency to unity-gain frequency, where PSRR is dominated by the open-loop gain of LDO. Within the frequency range of 450 kHz–1 MHz, the range of the PSRR results exceeds the unity-gain frequency, where feedback loop exerts a negligible effect; therefore, the output capacitor

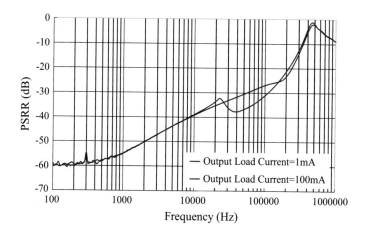

Fig. 1.4 Experimental results of PSRR for LDO at different output load currents (1 and 100 mA)

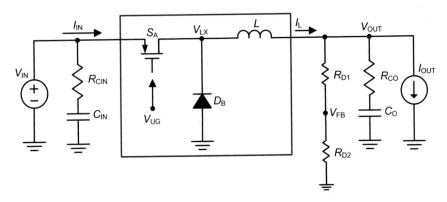

Fig. 1.5 Circuit diagram of an asynchronous buck converter

dominates along with the parasitic impedance from the input and output terminals. The gate driver's capability to drive power switch S_{MAIN} at the frequency range of 450 kHz–1 MHz is also an influencing factor.

The switching buck converter is the other type of DC–DC step-down converter. Figure 1.5 shows a circuit diagram of an asynchronous buck converter. A buck converter is applied to reduce DC voltage from the input voltage to the output voltage by using only power switches, main inductor L, input capacitors C_{IN}, and output capacitors C_O. Similar to the setup in any switching power converter, a power switch controls the transfer of energy. Inductors and capacitors are also used to store energy. Power switches are realized by using power semiconductor devices, such as metal–oxide–semiconductor field-effect transistor (MOSFET) S_A and diode D_B, which are controlled to turn on and turn off as required to regulate output voltage V_{OUT}. Usually, a P-channel MOSFET (PMOS) is used as MOSFET S_A instead of NMOS because if NMOS is employed as a high-side switch, driving it would be difficult because the gate and source are connected to the voltage supply. However, PMOS needs more internal circuits than NMOS. In the selection of diode D_B, three key specifications must be checked: reverse voltage, forward voltage drop, and forward current. The reverse voltage must be higher than the maximum voltage of voltage terminal V_{LX} to avoid damaging the diode. The forward voltage drop should be small. A large forward voltage drop can result in a large conduction loss. Meanwhile, the forward current must be larger than the maximum output load current; otherwise, it cannot provide sufficient energy to the output loading requirements. In addition, the inductor current ripple should be considered in the forward current rating for diode D_B. The forward current should be larger than the sum current between a half of inductor current ripple and the maximum output load current.

To achieve high load efficiency, diode D_B is usually replaced by n-channel MOSFET S_B as a low-side switch to increase load efficiency because NMOS can save more IC die size than PMOS. The total power loss in a DC–DC converter is significantly reduced by this replacement because the voltage drop in conducted

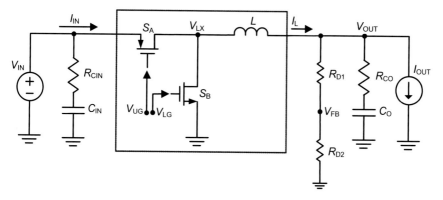

Fig. 1.6 Circuit diagram of a synchronous buck converter

NMOS is very low in comparison with that in a conducted diode (even the Schottky diode, which has a low forward voltage drop). A circuit diagram of a synchronous buck converter is shown in Fig. 1.6. The main advantage of a low-side switch is that the voltage drop across the low-side MOSFET can be lower than the voltage drop across the power diode of the asynchronous converter. When no change in the current level occurs, a low voltage drop translates to low power dissipation and high efficiency. For a synchronous buck converter, dead-time control must be implemented to prevent shoot-through current from flowing through the main power MOSFET during switching transitions by controlling the duty cycle of the MOSFET drivers. The high-side driver is not allowed to turn on until the gate drive voltage to the low-side MOSFET is low, and the low-side driver is not allowed to turn on until the voltage at the junction of the power MOSFET is low.

Figure 1.7 shows the control signals for buck converter operation. When high-side switch S_A is turned on by control signal V_{UG}, a path is provided for the DC input voltage to charge the inductor and supply the output loading requirements. Charging continues until feedback voltage V_{FB} reaches reference voltage V_{REF}. Then, the control part turns off the high-side switch to keep feedback voltage V_{FB} close to internal reference voltage V_{REF}. Therefore, no path exists to charge the inductor. Afterward, the inductor changes its voltage polarity, and the current flows in the same direction through low-side switch S_B, which is turned on by the control signal V_{LG} part with dead-time control. Discharging continues until feedback voltage V_{FB} is less than internal reference voltage V_{REF}. The control part again turns on high-side switch S_A to compensate for the output voltage drop. This cycle continues until complete regulation of output voltage is achieved.

This process is accomplished by sensing the output voltage of the circuit by means of a negative feedback loop that adjusts the duty cycle by Eq. (1.8) to control the on and off states of the MOSFET switches via control signals V_{UG} and V_{LG} under specified switching frequency F_S. In Eq. (1.8), T_{ON} is the time interval that power MOSFET S_A conducts (on state), T_{OFF} is the time interval that power MOSFET S_B conducts (off state), T_S is the period, and F_S is the switching

Fig. 1.7 Control signals for buck converter operation

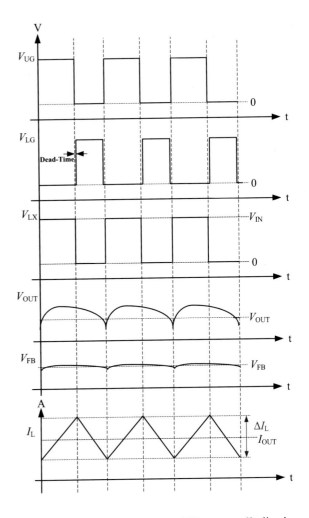

frequency. Switching frequency F_S of control signals V_{UG} and V_{LG} generally lies in the range of 100 kHz–2 MHz.

$$D = \frac{V_{OUT}}{V_{IN}} = \frac{T_{ON}}{T_S} = \frac{T_{ON}}{T_{ON} + T_{OFF}} = F_S \cdot T_{ON} \qquad (1.8)$$

The output voltage is also designed similar to that in an LDO circuit, as shown in Eq. (1.1). The inductor value and operating frequency determine inductor current ripple ΔI_L according to specific input and output voltages. Inductor current ripple ΔI_L has a positive correlation with input voltage. Thus, if the input voltage is increased, inductor current ripple ΔI_L can also be increased by Eq. (1.9). Having a low ΔI_L reduces not only ESR losses in the output capacitors but also the output voltage ripple. High frequency with a small ripple current can result in high-efficiency operations. In general, inductor current ripple ΔI_L is equal to 20% of the maximum load current. In poor conditions, the maximum input voltage can obtain the largest inductor current ripple ΔI_L, so the inductor value can be determined by Eq. (1.10).

$$\Delta I_L = \frac{V_{OUT}}{F_S \cdot L} \cdot \left(1 - \frac{V_{OUT}}{V_{IN}}\right) \tag{1.9}$$

$$L = \frac{V_{OUT}}{F_S \cdot \Delta I_{L\_MAX}} \cdot \left(1 - \frac{V_{OUT}}{V_{IN\_MAX}}\right) \tag{1.10}$$

Moreover, the selection of output capacitor C_O is determined by the required ESR R_{CO} to minimize the voltage ripple. The amount of bulk capacitance is also a key factor in C_O selection because it ensures that the control loop is stable. Output voltage ripple ΔV_{OUT} is determined by Eq. (1.11). ΔV_{OUT} also depends on inductor current ripple ΔI_L. Therefore, output voltage ripple ΔV_{OUT} has a positive correlation with input voltage because inductor current ripple ΔI_L has a positive correlation with input voltage. Increasing the input voltage would increase output voltage ripple ΔV_{OUT}. Multiple capacitors placed in parallel may need to meet ESR and RMS current handling requirements. In addition, ceramic capacitors are widely used as output capacitors because they are inexpensive and highly reliable and possess low ESR and small size. Equation (1.11) can be modified and rewritten as Eq. (1.12) without affecting ESR.

$$\Delta V_{OUT} \leq \Delta I_L \cdot \left(R_{CO} + \frac{1}{8 \cdot F_S \cdot C_O}\right) \tag{1.11}$$

$$\Delta V_{OUT} \leq \Delta I_L \cdot \frac{1}{8 \cdot F_S \cdot C_O} \tag{1.12}$$

For many power IC products and applications, the evaluation board size can be reduced by selecting a device in which the IC switching MOSFET elements are designed and embedded by a real implemented internal circuit. This method is needed to check the thermal of IC die temperature.

The output voltage of a buck converter must be smaller than the input voltage, so the output voltage is still smaller than the input voltage during the soft start time of a buck converter. A buck converter does not need the pre-charge function to regulate the output voltage. The soft start time is designed to reduce the inrush current at the input terminal. Figure 1.8 shows the experimental results of soft start with input voltage for buck converter. Input current I_{IN} is measured with an oscilloscope and is very small with a soft start time of 1.3 ms. In general, the current limit at the low-side MOSFET S_B function can limit the input energy to the output terminal. Hence, input current I_{IN} should be reduced at the soft start time. In addition, internal reference voltage V_{REF} is controlled with a slow slew rate. These two means can reduce the inrush current at the input terminal.

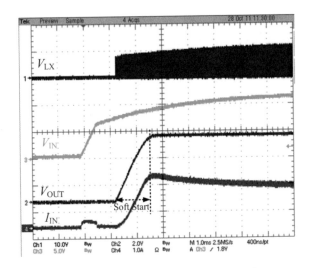

Fig. 1.8 Experimental results of soft start with input voltage for buck converter

Figure 1.9 shows a circuit diagram of the synchronous boost converter. A boost converter is applied to increase the DC voltage from the input voltage to the output voltage by using only power switches, main inductor L, input capacitors C_{IN}, and output capacitors C_O. As in any switching power converter, a power switch controls the transfer of energy. Inductors and capacitors are also used to store energy. Power switches are realized by using power semiconductor devices, such as MOSFET S_A and MOSFET S_B, which are controlled to turn on and turn off as required to regulate output voltage V_{OUT}. Usually, PMOS is preferred to be used as MOSFET S_A instead of NMOS, similar to buck converter. The n-channel MOSFET S_B as a low-side switch increases the load efficiency. The main advantage of a low-side switch is that the voltage drop across the low-side MOSFET can be lower than the

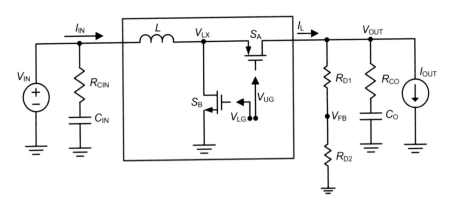

Fig. 1.9 Circuit diagram of the synchronous boost converter

Fig. 1.10 Control signals in the boost converter operation

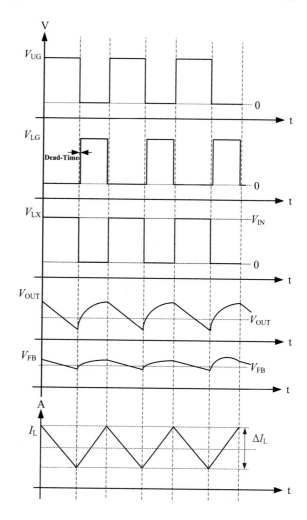

voltage drop across the power diode of the asynchronous converter. When no change in the current level occurs, a low voltage drop translates to low power dissipation and high efficiency.

Figure 1.10 shows the control signals in the boost converter operation. When low-side switch S_B is turned on by control signal V_{LG}, a path is provided for the DC input voltage to charge the inductor and cut the supply of the output loading requirement. Charging continues until feedback voltage V_{FB} reaches reference voltage V_{REF}. Then, the control part turns off the low-side switch to keep feedback voltage V_{FB} close to internal reference voltage V_{REF}. Therefore, no path exists to charge the inductor. The inductor then changes its voltage polarity, and the current flows in the same direction through high-side switch S_A, which is turned on by the control signal V_{UG} part with dead-time control. Discharging continues until feedback voltage V_{FB} is less than internal reference voltage V_{REF}. The control part again turns on low-side switch S_B to compensate for the output voltage drop. This cycle continues until complete regulation of output voltage is achieved.

This process is accomplished by sensing the output voltage of the circuit by means of a negative feedback loop that adjusts the duty cycle by Eq. (1.13) to control the on and off states of the MOSFET switches via control signals V_{UG} and V_{LG} under specified switching frequency F_S. In Eq. (1.13), T_{ON} is the time interval that power MOSFET S_B conducts (on state), T_{OFF} is the time interval that power MOSFET S_A conducts (off state), T_S is the period, and F_S is the switching frequency.

$$D = \frac{V_{OUT} - V_{IN}}{V_{OUT}} = \frac{T_{ON}}{T_S} = \frac{T_{ON}}{T_{ON} + T_{OFF}} = F_S \cdot T_{ON} \tag{1.13}$$

The output voltage of the boost converter is also designed similar to that of buck converter by using Eq. (1.1). The inductance depends on the maximum input current because the inductor current is equal to the input current, as shown in Eq. (1.14). As a general rule, the inductor current ripple is designed around 20% to 40% of the maximum input current. Assuming that the inductor current ripple is equal to 40% of the input current, it can be obtained with Eq. (1.15). To select a suitable inductor, the inductor saturated current should be larger than the sum current between a half of inductor current ripple and the maximum inductor current. The inductance should be determined, as shown in Eq. (1.16), where F_S is the switching frequency. To consider system performance, a shielded inductor is preferred to avoid the EMI issue.

$$I_{L\_MAX} = \frac{V_{OUT} \cdot I_{OUT\_MAX}}{\eta \cdot V_{IN}} \tag{1.14}$$

$$\Delta I_{L\_RIPPLE} = 40\% \cdot I_{L\_MAX} \tag{1.15}$$

$$L = \frac{\eta \cdot V_{IN}^2 \cdot (V_{OUT} - V_{IN})}{40\% \cdot V_{OUT}^2 \cdot I_{OUT\_MAX} \cdot F_S} \tag{1.16}$$

The selection of output capacitor C_O is determined by the required ESR R_{CO} to minimize the voltage ripple. The amount of bulk capacitance is also a key factor in C_O selection to ensure that the control loop is stable. Output voltage ripple ΔV_{OUT} is determined by Eq. (1.17). ΔV_{OUT} also depends on output load current I_{OUT} and inductor current I_L. Multiple capacitors placed in parallel may need to meet ESR and root-mean-square (RMS) current handling requirements. Ceramic capacitors are widely used as output capacitors because they are inexpensive and highly reliable and possess low ESR and small size. Equation (1.17) can be modified and rewritten as Eq. (1.18) without affecting ESR.

$$\Delta V_{OUT} = \Delta I_L \cdot R_{CO} + \frac{D \cdot I_{OUT}}{\eta \cdot F_S \cdot C_O} \qquad (1.17)$$

$$\Delta V_{OUT} = \frac{D \cdot I_{OUT}}{\eta \cdot F_S \cdot C_O} \qquad (1.18)$$

The output voltage of the boost converter must be larger than the input voltage, so the output voltage may be larger than the input voltage during the soft start time of the boost converter. However, the output voltage is zero in the initial condition, so most IC products for the boost converter have a pre-charge function. This function can achieve an output voltage that is equal to the input voltage. The output voltage can then be regulated to meet the user design rating. The boost converter not only has a soft start time similar to buck converter; it also has a pre-charge time.

Soft start time and pre-charge time are designed to reduce the inrush current at the input terminal. Figure 1.11 shows the experimental results of soft start with input voltage for the boost converter. The output voltage is equal to 3.3 V, similar to the input voltage at pre-charge time. The pre-charge time is close to 2 ms. Based on the pre-charge implemented circuit, three methods can be adopted to achieve an output voltage that is equal to the input voltage with a low inrush current at the input terminal. These methods should be avoided for the high-side MOSFET S_A to fully turn on. The first method senses the high-side MOSFET S_A to clamp its current and reduce inductor current I_L. The second method produces the other couple control signals of V_{UG} and V_{LG} to deliver input energy to the output terminal slowly. The last method is similar to the second method and reduces the switching frequency for the other couple control signals of V_{UG} and V_{LG} to deliver input energy to the output terminal slowly.

Inductor current I_L is equal to input current I_{IN} and is measured with an oscilloscope. Inductor current I_L very small with a soft start time of 1.3 ms. In general, the current limit at the low-side MOSFET S_B function can limit the input energy to

Fig. 1.11 Experimental results of soft start with input voltage for boost converter

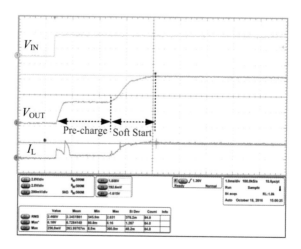

the output terminal. Hence, inductor current I_L should be reduced at soft start time. In addition, internal reference voltage V_{REF} can be controlled with a slow slew rate. These two means can reduce the inrush current at the input terminal.

1.2   Major Control Modes for Power Converters

The major control modes according to the couple control signals of V_{UG} and V_{LG} can be classified as pulse width modulation (PWM) control mode and pulse frequency modulation (PFM) control mode [1–11]. The PWM control mode can generate couple control signals of V_{UG} and V_{LG} for a fixed switching frequency. The switching frequency is a constant cycle by cycle at the steady status. Figure 1.12 shows the PWM control mode to generate couple control signals for V_{UG} and V_{LG}. The PWM control mode is widely used to control the duty cycle regulating the output voltage to meet the user design. For power IC products, the current-mode and voltage-mode control circuits belong to the PWM control mode because both circuits use a clock signal V_{CLOCK} to make a fixed switching frequency. The pulse of V_{CLOCK} is used to trigger the SR flip-flop, and the input terminal begins to deliver energy to the output terminal until the energy is sufficient at the output terminal and feedback voltage V_{FB} reaches reference voltage $V_{REF.}$ The control circuit then generates another pulse to reset the SR flip-flop.

The PFM control mode can generate the couple control signals of V_{UG} and V_{LG} for a variable switching frequency, so the on-time width or off-time width is the same width cycle by cycle at the steady status. When the output terminal needs

Fig. 1.12 PWM control mode to generate the couple control signals for V_{UG} and V_{LG}

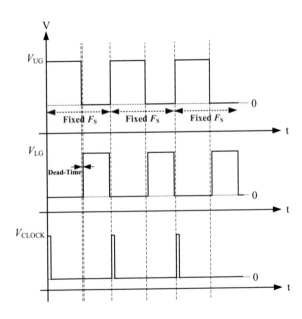

more energy from the input terminal, the PFM control mode may generate one control signal V_{TRIG} to trigger the on-time generation circuit or off-time generation circuit. For power IC products, the on-time generation circuit or off-time generation circuit is designed and implemented by Eq. (1.19).

$$T_{ON} = \frac{C \cdot V_{REFA}}{I} \tag{1.19}$$

Figure 1.13 shows a circuit diagram of the on-time generation circuit with SR flip-flop for PFM control mode. The charged current I_{CHARGE} is calculated by Eq. (1.20). I_{CHARGE} can charge capacitor C_1 when control signal V_{LG} turns off. Then, voltage V_{CHARGE} is increased linearly. When V_{CHARGE} is larger than reference voltage V_{REF2} and the output signal V_{CMP} of the comparator is changed from low level to high level, the control signals of V_{UG} and V_{LG} can be used to drive and control power MOSFETs S_A and S_B.

$$T_{ON} = \frac{C \cdot V_{REF2}}{\left(\frac{V_{REF1}}{R_1}\right) \cdot G_1} \tag{1.20}$$

Figure 1.14 shows a PFM control mode to generate the couple control signals for V_{UG} and V_{LG}. The PFM control mode is widely used to control the variable switching frequency regulating the output voltage to meet the user design. A major difference of PFM control is not having a clock signal V_{CLOCK}. Given that PFM is not a constant frequency, it does not need V_{CLOCK} to keep the same switching frequency. If the output terminal needs more energy to keep the output voltage and feedback voltage V_{FB} is less than internal reference voltage V_{REF}, control signal V_{TRIG} should be triggered from low level to high level. The pulse of clock signal V_{TRIG} is used to trigger SR flip-flop, and the input terminal begins to deliver energy to the output terminal until the capacitor's voltage V_{CHARGE} is larger than internal reference voltage V_{REF2}. The on-time generation circuit then controls the output terminal of comparator V_{CMP} from low level to high level to reset SR flip-flop.

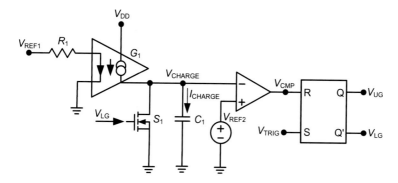

Fig. 1.13 Circuit diagram of on-time generation circuit with SR flip-flop for PFM control mode

Fig. 1.14 PFM control mode to generate the couple control signals for V_{UG} and V_{LG}

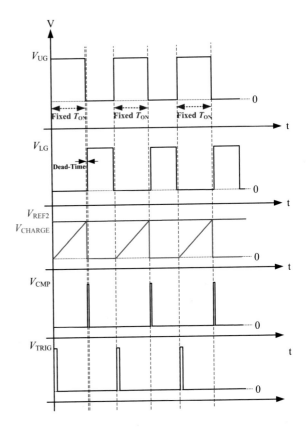

For power IC products, the conventional constant on-time control circuit and adaptive on-time control circuit belong to the PFM control mode. The conventional constant on-time control circuit can generate the same on-time width regardless of the input or output voltage variation. The switching frequency should be increased if the output terminal is under a heavy load and the on-time generation circuit generates more control pulses to deliver more energy from the input terminal to the output terminal. If the output terminal is under high output voltage and the on-time generation circuit generates more control pulses to deliver more energy from the input terminal to the output terminal, the switching frequency should also be increased. In addition, the switching frequency of conventional constant on-time control circuit depends on the output load current and output voltage.

Meanwhile, the on-time generation circuit of adaptive on-time control circuit is used to sample the input and output voltages. The adaptive on-time control circuit can be classified as constant frequency on-time and constant current ripple on-time according to the connection form. Constant frequency on-time control circuit is different from conventional constant on-time (COT) control circuit because it prevents the generation of the same on-time width at a high output voltage or under heavy load. Constant current ripple on-time control circuit is also different from conventional COT control circuit because it prevents the generation of the same on-time width at the voltage drop between the input and output voltages. Therefore, even when the voltage drop between the input and output voltages is changed, the inductor current ripple maintains a constant value.

1.3   Load Transient Response and Load Regulation

Switching power supply with power IC is commonly used for fast load transient response because it can improve the load transient response to reduce output capacitance, especially in CPU and load applications with high current slew rate. Most modern consumer electronic products contain embedded CPU functions, wireless connectivity, and media players. When these electronic products are used and operated, the output loading requirements are almost like a pulsating load. A pulsating load generally involves a fast slew rate of rising and falling time. When consumer electronic products need more energy and the output load current is changed from light to heavy, it is called load droop. If the energy is sufficient to supply consumer electronic products and the output load current is also changed from heavy to light, it is called load release. Load release and load droop belong to load transient.

For power IC products, load transient is the most important measured item. The experimental results of load transient are known as IC performance and characteristic. The load transient response is a quick way to determine the power converter control loop response. The power converter control loop can determine the power converter regulation speed and reveal control loop stability problems. The load transient response can address the steady-status error further to determine the load regulation performance.

Figure 1.15 shows the experimental results for the load transient without a steady-status error. The operation conditions are as follows: input voltage of 12 V, output voltage of 1.05 V, switching frequency of 650 kHz, the output capacitor is 22 μF, and the inductor is 1.4 μH. The red-colored waveform represents the output voltage, and the pink-colored waveform represents the output load current. According to the experimental results, the output voltage suffers a voltage drop at the load droop. At the same time, the power IC needs to convert the input power source to deliver more energy to the output loading requirements to minimize the output voltage drop. Once the input power source delivers sufficient energy to the output loading requirements at load release, the output voltage suffers an overshoot. The extra pulse at load release must be avoided because a large overshoot results in damage to consumer electronic products.

The experimental results of load transient show the same voltage level of output voltage under heavy and light loads, so this power converter has no steady-status error and demonstrates good load regulation performance.

Figure 1.16 shows the experimental results for load regulation without a steady-status error at an output voltage of 1.05 V under different input voltages of 5,

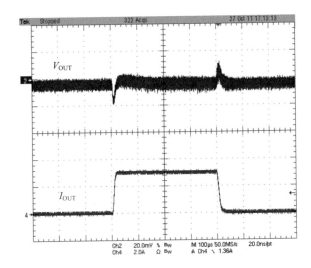

Fig. 1.15 The experimental results for the load transient without a steady-status error

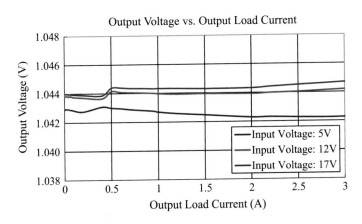

Fig. 1.16 The experimental results for the load regulation without a steady-status error at an output voltage of 1.05 V under different input voltages of 5, 12, and 17 V

12, and 17 V. The specification of load regulation is generally defined as +1%/−1% of the output voltage. Based on this operation condition, the output voltage should range from 1.0395 V to 1.0605 V. The experimental results meet this specification of load regulation and can prove that the output load current does not affect the output voltage regulation. The experimental results are also measured under different input voltages of 5, 12, and 17 V. The experimental results prove that the input voltage does not affect the output voltage regulation. For these reasons, this power converter with power IC is suitable for applications with a fixed output voltage under different output load currents and input voltages.

If the system control loop response is fast, the output voltage may suffer from a low voltage drop. If the system control loop response is too fast, then the power converter may suffer from noise jitter, subharmonic, or even result in an unstable system. If the system control loop response is slow, the output voltage may suffer from a large voltage drop. A large voltage drop causes consumer electronic products to suffer from damage or failure; hence, the design of the system control loop response is a trade-off. The system control loop response is an optimal design based on its operation conditions and must cover the worst operation conditions.

Most IC products have no steady-status error design because the electronic device or equipment cannot accept a large variation of the output voltage. A large steady-status error cannot meet the specification of load regulation. CPU requires a unique power delivery and allows the supply voltage to have a large steady-status error. CPU generally uses adaptive voltage position (AVP) to achieve the steady-status error. Output voltage has an inverse correlation with output load current for CPU with AVP [27–33]. If the output load current is changed from light to heavy, the output voltage for delivery to the CPU will be reduced. Output impedance may behave as an equivalent resistant R_{LL} by Eq. (1.21), as shown in Fig. 1.17. Figure 1.17 shows the voltage identification (VID) V_{VID} voltage by a real IC implemented circuit, and the user should base on the VID code table to determine the V_{VID} voltage by the I²C interface. If the CPU's load current is zero, output voltage V_{OUT} is close to V_{VID} voltage. The resistor R_{LL} is generally designed with external components, and the V_{VID} voltage is equal to the sum voltage between voltage droop V_{DROOP} and output voltage V_{OUT}. The load transient response occurs in the CPU and should not require much energy to keep and regulate the same output voltage; hence, the steady-status error can save on energy and capacitors.

$$V_{OUT} = V_{VID} - I_{OUT} \times R_{LL} \qquad (1.21)$$

Figure 1.18 shows the experimental results for load transient with a steady-status error using AVP. The operation conditions are as follows: input voltage of 12 V, output voltage of 1.1 V, switching frequency of 350 kHz, the output capacitors consist of three OS-CON capacitors (560 μF) and 22 ceramic capacitors (22 μF), and the inductor is 360 nH for each phase. The blue-colored waveform represents the output voltage, and the black-colored waveform represents the output load current. According to the experimental results, the output voltage suffers from a voltage drop at load droop. At the same time, the power IC needs to convert the

Fig. 1.17 Equivalent circuit for V_{VID} voltage by a real IC implemented circuit with AVP

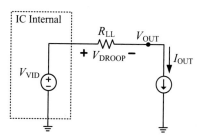

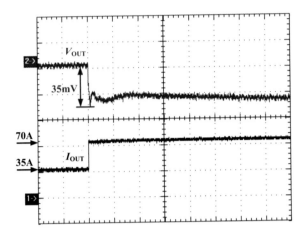

Fig. 1.18 The experimental results for the load transient with a steady-status error using AVP

input power source to deliver more energy to the output loading requirements to minimize the output voltage drop. The output voltage can be controlled and regulated at a different voltage level according to the output load current. The output load current is changed from 35 to 70 A, and the output voltage may be changed from 1.1 to 1.065 V so as not to regulate the same output voltage of 1.1 V. Based on this experimental result, resistor R_{LL} is defined as 1 mΩ, and voltage droop V_{DROOP} is equal to 35 mV.

The experimental results of the load transient show a different voltage level of the output voltage at heavy and light loads, so this power converter has a steady-status error using AVP and shows poor load regulation performance.

Figure 1.19 shows the experimental results for load regulation with a steady-status error at an output voltage of 1.1 V and input voltage of 12 V. The

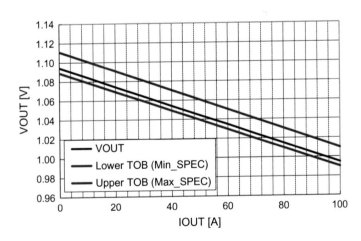

Fig. 1.19 The experimental results for load regulation with a steady-status error at an output voltage of 1.1 V and input voltage of 12 V

specification of load regulation generally defines the upper tolerance of the band (TOB) as the maximum voltage and lower TOB as the minimum voltage of the output voltage. Based on this operation condition, the output voltage should be within upper and lower TOB limits. The experimental results meet this specification of load regulation and can prove that the output load current directly affects the output voltage regulation. For these reasons, this power converter with power IC is suitable for applications with a steady-status error using AVP, such as CPU load delivery.

Figure 1.20 shows a comparison of PWM and PFM control modes at load droop under the same operation condition [34–36]. Comparing the load transient between PWM and PFM control modes is difficult because two similar topologies for PWM and PFM control modes must be determined. Interestingly, the current-mode control circuit belongs to the PWM control mode, which is very similar to the current-mode adaptive on-time control circuit. The current-mode adaptive on-time control circuit belongs to the PFM control mode. The on-time generator of the current-mode adaptive on-time control circuit is significantly different from that of the current-mode control circuit. The current-mode adaptive on-time control circuit is highly similar to the current-mode control circuit. Given the different on-time generator, it is easy to determine the load transient response between PWM and PFM control modes at load droop. These simulation results use the same operation

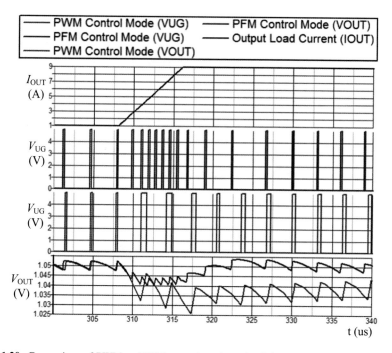

Fig. 1.20 Comparison of PWM and PFM control modes at load droop under the same operation condition

conditions to compare PWM and PFM control modes at an output voltage V_{OUT} of 1.05 V, input voltage V_{IN} of 19 V, switching frequency F_S of 320 kHz, the output capacitors C_O of two PSCAP is 330 µF, the inductor L is 2.2 µH, and light load of 1 A to heavy load of 9 A within 8 µs. The PFM control mode can generate multiple pulses with a minimum off-time mechanism to prevent V_{OUT} from decreasing significantly. The PFM control mode is useful for the reduction of V_{OUT} peak-to-peak voltage at load droop. The PWM control mode can simply increase its PWM pulse width with the same switching frequency at the droop. Comparison of PFM and PWM control modes at the droop shows that the former can generate more PWM pulses than the latter. Therefore, the PFM control mode can achieve a faster transient response than the PWM control mode. In addition, the PFM control mode can achieve shorter settling time than the PWM control mode.

1.4 Efficiency

The efficiency of DC–DC power converters with ICs is a critical criterion in different industries, such as consumer electronics, automotive, telecommunication, networking, and medicine. The efficiency of the selected power solutions relates to system power loss and the thermal performance of ICs, PCBs, and other components, which determines the power effectiveness. High-efficiency DC–DC power converters with ICs result in minimal heat dissipation, which reduces system cost and the size of elements, such as heat sinks, fans, and their assembly. For example, in a battery-operated system, less power loss means that the device can use the same battery for a longer run time because the device pulls less current from the battery.

An ideal DC–DC power converter has no loss on the components and power switch; it has no switching and conduction losses. An ideal power switch implies zero losses, thus offering 100% efficiency. However, the components are not ideal, as illustrated in the following examples.

In general, to consider the various factors that contribute to efficiency, the focus should be on buck converter, which is the most popular DC–DC power converter applied in consumer electronics, automotive, telecommunication, networking, and medical applications. A high-efficiency buck converter should achieve low power dissipation. Figure 1.21 shows the power dissipation in the inductor and MosFETs

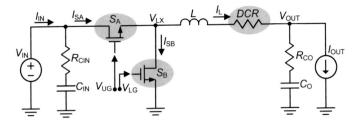

Fig. 1.21 The power dissipation in the inductor and MosFETs for buck converter

Fig. 1.22 The inductor current I_L for buck converter

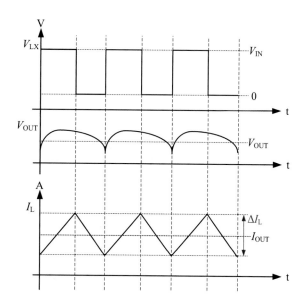

for buck converter. The three main causes of power dissipation in a buck converter are inductor conduction loss, MOSFET conduction loss, and MOSFET switching loss, which occur in the components when buck converter operates in continuous conduction mode (CCM), fixed switching frequency, fixed input voltage, and fixed output voltage [37–46].

Figure 1.22 shows the inductor current for buck converter. DCR is the DC resistor of the inductor. The power dissipation in DCR addresses the inductor conduction loss in Eq. (1.22). The RMS inductor current is shown in Eq. (1.23). The RMS inductor current can be written to approximate the output load current as Eq. (1.24) because inductor current ripple ΔI_L is not a major influencing factor of RMS inductor current. The inductor selection and output load current are usually decided by the customers or users, so improving inductor conduction loss is difficult.

$$P_{DCR} = I_{L\_RMS}^2 \times DCR \tag{1.22}$$

$$I_{L\_RMS}^2 = I_{OUT}^2 + \frac{\Delta I_L^2}{12} \tag{1.23}$$

$$I_{L\_RMS}^2 \approx I_{OUT}^2 \tag{1.24}$$

MOSFET conduction loss means power switch S_A is turned on and experiences power dissipation. Power switch S_B is turned on and also experiences power dissipation. When any power switch is turned on, this power switch can be modeled as resistor R_{ON}. The conduction loss of power switch S_A can be calculated by Eqs. (1.25)–(1.26). The conduction loss of power switch S_A occurs when power

Fig. 1.23 The current I_{SA} of power switch S_A for buck converter

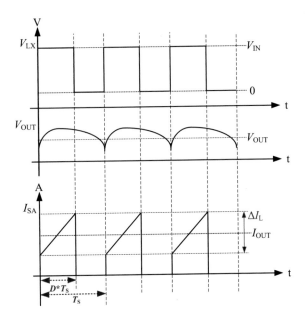

switch S_A is turned on during $D*T_S$, so current I_{SA} goes through resistor R_{SA_ON} as shown in Fig. 1.23.

$$P_{SA\_CON} = I_{SA\_RMS}^2 \times R_{SA\_ON} \tag{1.25}$$

$$P_{SA\_CON} = \frac{V_{OUT}}{V_{IN}} \times \left(I_{OUT}^2 + \frac{\Delta I_L^2}{12} \right) \times R_{SA\_ON} \tag{1.26}$$

The conduction loss of power switch S_B can be calculated by Eqs. (1.27)–(1.28). The conduction loss of power switch S_B occurs when power switch S_B is turned on during $D'*T_S$, so current I_{SB} goes through resistor R_{SB_ON} as shown in Fig. 1.24.

$$P_{SB\_CON} = I_{SB\_RMS}^2 \times R_{SB\_ON} \tag{1.27}$$

$$P_{SB\_CON} = \left(1 - \frac{V_{OUT}}{V_{IN}} \right) \times \left(I_{OUT}^2 + \frac{\Delta I_L^2}{12} \right) \times R_{SB\_ON} \tag{1.28}$$

The overall conduction loss of MOSFETs is equal to the sum of power dissipation P_{SA_CON} and power dissipation P_{SB_CON}. The conduction loss of MOSFETs depends on the output load current. For this reason, if the output load current is heavy load, the conduction loss of MOSFETs should be increased. The efficiency under heavy load will improve if the conduction loss of MOSFETs is reduced. One means to reduce the conduction loss of MOSFETs is to reduce their turned-on resistor R_{ON}. However, low resistor R_{ON} needs more IC die size, which depends on

Fig. 1.24 The current I_{SB} of power switch S_B for buck convert

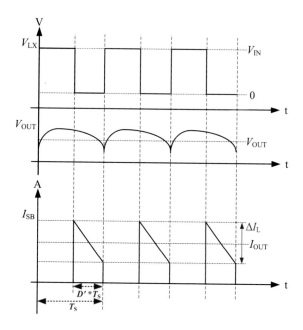

the IC cost. Meanwhile, a large conduction loss results in a large turned-on resistor R_{ON}, which may increase the IC junction temperature. The IC must be checked to prevent it from suffering from a thermal shutdown.

MOSFETs have a very short switching time, so the switching loss of MOSFETs originates from dynamic drain-to-source voltage V_{DS} and drain current I_D that the MOSFETs must handle during the time it takes to turn on or off.

Figure 1.25 shows the level shift driver circuit that drives power switch S_A. This level shift driver circuit is applied to change the power domain from IC internal power to external input voltage V_{IN}. This function of the level shift driver circuit can provide strong power to control power switch S_A. Figure 1.25 shows the equivalent circuit of power switch S_A with internal parasitic capacitors, such as C_{GD}, C_{GS}, and C_{DS}.

When power switch S_A is turned on, internal parasitic capacitor C_{GS} must charge until the voltage of capacitor C_{GS} is equal to threshold voltage V_{TH}. When the voltage of capacitor C_{GS} is larger than threshold voltage V_{TH}, current I_{DS} of power switch S_A is increased. Current I_{DS} increases to equal output load current I_{OUT} until the voltage of capacitor C_{GS} equals the Miller plateau voltage V_{PLAT} within T_1 time. At this moment, the power delivery V_{IN} of the level shift driver circuit charges the Miller parasitic capacitor C_{GD} so that the voltage of capacitor C_{GS} can be maintained for a while. At the same time, voltage V_{LX} is increased to equal input voltage V_{IN}, and voltage V_{DS} is reduced to zero within T_2 time (Fig. 1.26). In Fig. 1.26, the turn on switching loss P_{SA_SWON} of power switch S_A is equal to the power dissipation in the area during T_1 and T_2 time, as shown in Eqs. (1.29)–(1.32).

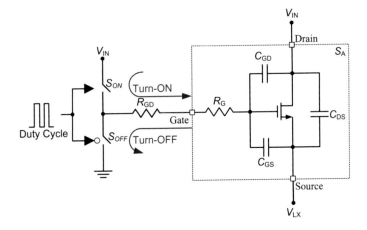

Fig. 1.25 The level shift driver circuit that drives power switch S_A

Fig. 1.26 Turn on switching loss P_{SA_SWON} of power switch S_A in the area during T_1 time and T_2 time

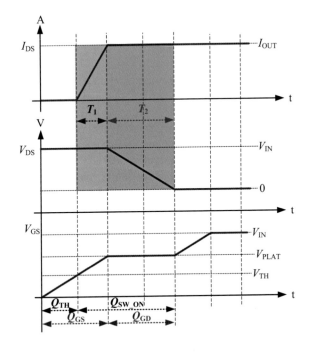

$$P_{SA\_SWON} = F_S \times V_{DS} \times I_{DS} \times \frac{T_1 + T_2}{2} = F_S \times V_{IN} \times I_{OUT} \times \frac{T_{SA\_SWON}}{2} \quad (1.29)$$

$$T_{SA\_SWON} = \frac{Q_{SW\_ON}}{I_{SA\_SWON}} \quad (1.30)$$

$$I_{\text{SA\_SWON}} = \frac{V_{\text{IN}} - V_{\text{PLAT}}}{R_{\text{SON\_ON}} + R_{\text{GD}} + R_{\text{G}}} \tag{1.31}$$

$$P_{\text{SA\_SWON}} = F_{\text{S}} \times V_{\text{IN}} \times I_{\text{OUT}} \times \frac{Q_{\text{SW\_ON}}}{2 \times I_{\text{SA\_SWON}}} \tag{1.32}$$

When power switch S_{A} is turned off, internal parasitic capacitor C_{GS} must discharge until the voltage of capacitor C_{GS} is equal to Miller plateau voltage V_{PLAT}. At this moment, Miller parasitic capacitor C_{GD} is also discharged to equal Miller plateau voltage V_{PLAT} so that the voltage of capacitor C_{GS} can be maintained for a while. At the same time, voltage V_{LX} is reduced to zero, and voltage V_{DS} is increased to input voltage V_{IN} within T_3 time (Fig. 1.27). When the voltage of capacitor C_{GS} is smaller than Miller plateau voltage V_{PLAT}, current I_{DS} is reduced to zero until the voltage of capacitor C_{GS} is equal to threshold voltage V_{TH} within T_4 time (Fig. 1.27). In Fig. 1.27, the turn off switching loss $P_{\text{SA\_SWOFF}}$ of power switch S_{A} is equal to the power dissipation in the area during T_1 and T_2 time, as shown in Eqs. (1.33)–(1.36).

$$P_{\text{SA\_SWOFF}} = F_{\text{S}} \times V_{\text{DS}} \times I_{\text{DS}} \times \frac{T_3 + T_4}{2} = F_{\text{S}} \times V_{\text{IN}} \times I_{\text{OUT}} \times \frac{T_{\text{SA\_SWOFF}}}{2} \tag{1.33}$$

$$T_{\text{SA\_SWOFF}} = \frac{Q_{\text{SW\_OFF}}}{I_{\text{SA\_SWOFF}}} \tag{1.34}$$

Fig. 1.27 Turn off switching loss $P_{\text{SA\_SWOFF}}$ of the power switch S_{A} in the area during T_3 time and T_4 time

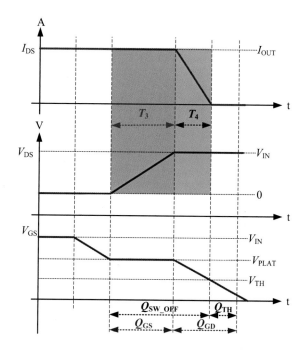

$$I_{\text{SA\_SWOFF}} = \frac{V_{\text{PLAT}} - 0}{R_{\text{SOFF\_ON}} + R_{\text{GD}} + R_{\text{G}}} \tag{1.35}$$

$$P_{\text{SA\_SWOFF}} = F_{\text{S}} \times V_{\text{IN}} \times I_{\text{OUT}} \times \frac{Q_{\text{SW\_OFF}}}{2 \times I_{\text{SA\_SWOFF}}} \tag{1.36}$$

Finally, the switching loss $S_{\text{A\_SW}}$ of power switch S_{A} can be computed as the difference between turn on switching loss $P_{\text{SA\_SWON}}$ and turn off switching loss $P_{\text{SA\_SWOFF}}$ by using Eq. (1.37).

$$P_{\text{SA\_SW}} = P_{\text{SA\_SWON}} + P_{\text{SA\_SWOFF}} \tag{1.37}$$

Figure 1.28 shows the level shift driver circuit that drives power switch S_{B}. This level shift driver circuit is applied to change the power domain from IC internal power to external input voltage V_{IN}. Power switch S_{B} has internal parasitic capacitors, such as C_{GD}, C_{GS}, and C_{DS}. Power switch S_{B} is turned on by the voltage of capacitor C_{GS}, and voltage V_{DS} is equal to the voltage drop of the body diode, which is generally smaller than 0.7 V. For this reason, the turn on switching loss $P_{\text{SB\_SWON}}$ of power switch S_{B} can be negligible. Meanwhile, power switch S_{B} is turned off by the voltage of capacitor C_{GS}, and the output load current I_{OUT} continues running in the same direction through the body diode. Thus, voltage V_{DS} is also small. For the same reason, the turn on switching loss $P_{\text{SB\_SWON}}$ of power switch S_{B} can be negligible. Accordingly, the switching loss $S_{\text{B\_SW}}$ of power switch S_{B} is minimal.

The switching loss of MOSFETs is equal to the sum of power dissipation $P_{\text{SA\_SW}}$ and power dissipation $P_{\text{SB\_SW}}$. Thus, the switching loss of MOSFETs approximates power dissipation $P_{\text{SA\_SW}}$. The switching loss of MOSFETs generally occurs at a short time during the period when the power switches are turned on

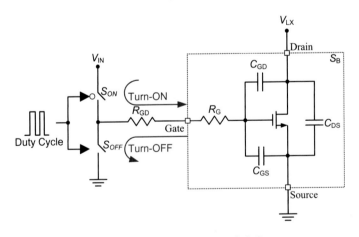

Fig. 1.28 The level shift driver circuit that drives power switch S_{B}

and turned off. Hence, it exerts no effect when the output load current is heavy. Moreover, when the output load current is light and the switching loss of MOSFETs is continuous, load efficiency is affected. For this reason, the output load current is light, and the switching loss of MOSFETs should be maintained. Therefore, the switching loss of MOSFETs should be reduced to increase efficiency at light loads.

The PFM control mode is widely implemented to reduce the switching loss of MOSFETs because the PFM control mode does not generate the same number of control pulses like the PWM control mode does. At a light load, the PWM control mode should force the duty cycle generation at the same switching frequency, so the PWM control mode may result in a larger switching loss than the PFM control mode. Figure 1.29 shows the control signals of the PWM control mode for buck converter. For a brief analysis, Fig. 1.29 ignores the effects of output voltage ripple, dead-time consideration, voltage drop of power switch S_A, and voltage drop of power switch S_B. The minimum inductor current I_{L_MIN} is a negative current, so the PWM control mode allows a negative inductor current to waste energy. However, the zero-current detection (ZCD) function [47–51] is widely used to prevent the inductor current I_L from becoming a negative current to achieve improved

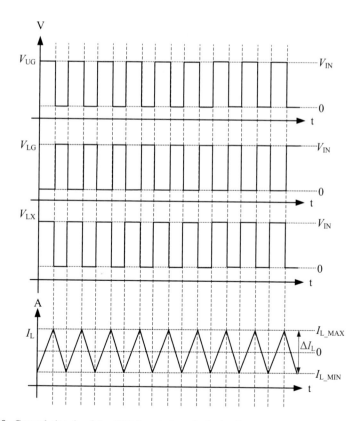

Fig. 1.29 Control signals of the PWM control mode for buck converter

efficiency at a light load. The ZCD function is usually implemented with the PWM control mode or PFM control mode to improve efficiency at a light load.

The ZCD function can detect inductor current I_L or voltage V_{LX} to control power switch S_B. If the ZCD function detects inductor current I_L, the ZCD function will change the voltage signal V_{ZCD} from high to low level to turn off power switch S_B when inductor current I_L is less than or equal to zero. The ZCD function can also detect voltage V_{LX}. When voltage V_{LX} is greater than or equal to zero, the ZCD function changes the voltage signal V_{ZCD} from high to low level to turn off power switch S_B. At this moment, inductor current I_L goes through the body diode of power switch S_B, and inductor current I_L equals zero quickly. Thus, voltage V_{LX} is equal to output voltage V_{OUT} and not equal to zero voltage, such as the PWM control mode without the ZCD function.

Figure 1.30 shows the control signals of the PWM control mode with the ZCD function for buck converter. For a brief analysis, Fig. 1.30 ignores the effects of

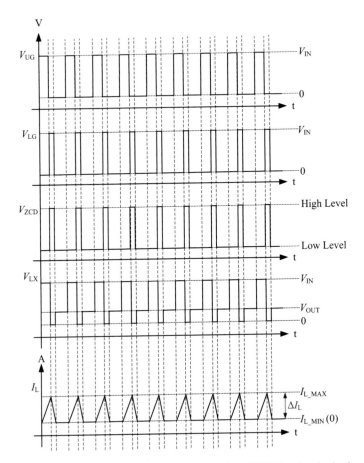

Fig. 1.30 Control signals of the PWM control mode with the ZCD function for buck converter

output voltage ripple, dead-time consideration, voltage drop of power switch S_A, and voltage drop of power switch S_B. The minimum inductor current I_{L_MIN} is equal to zero, so the PWM control mode with the ZCD function has no negative inductor current to waste energy. The width of voltage V_{UG} for the PWM control mode with the ZCD function is according to the output load current, not to the ratio of the output voltage to the input voltage. Thus, the width of voltage V_{UG} for the PWM control mode with the ZCD function is shorter than the duty cycle of the PWM control mode. Once the voltage signal V_{ZCD} is changed from high to low level to turn off power switch S_B, the width of voltage V_{LG} for the PWM control mode with the ZCD function is according voltage signal V_{ZCD}. The inductor current ripple ΔI_L of the PWM control mode is larger than that of the PWM control mode with the ZCD function because the inductor current I_L of the PWM control mode with the ZCD function has no negative current. Hence, the maximum inductor current I_{L_MAX} should be less than the PWM control mode without the ZCD function.

In general, the PFM control mode with the ZCD function not only prevents the negative inductor current from wasting energy for power ICs but can also significantly reduce the number of control pulses to increase light load efficiency. Figure 1.31 shows the control signals of the PFM control mode with the ZCD function for buck converter. For a brief analysis, Fig. 1.31 ignores the effects of output voltage ripple, dead-time consideration, voltage drop of power switch S_A, and voltage drop of power switch S_B. The minimum inductor current I_{L_MIN} is equal to zero, similar to the PWM control mode with the ZCD function. The width of voltage V_{UG} for the PWM control mode with the ZCD function is according to the ratio of output voltage to input voltage, so the width of voltage V_{UG} for the PFM control mode with the ZCD function is equal to the duty cycle of the PWM control mode. For this reason, the inductor current ripple ΔI_L of the PFM control mode with the ZCD function is larger than that of the PWM control mode. The inductor current ripple ΔI_L of the PFM control mode with the ZCD function is equal to that of the PWM control mode because the width of voltage V_{UG} for the PFM control mode with the ZCD function is similar to that of the PWM control mode. Therefore, the inductor current ripple ΔI_L of the PFM control mode with the ZCD function is similar to that of the PWM control mode. However, the minimum inductor current I_{L_MIN} for the PWM control mode is negative, so the PWM control mode cannot save extra energy like the PFM control mode with the ZCD function does.

Once voltage signal V_{ZCD} is changed from high to low level to turn off power switch S_B, the width of voltage V_{LG} for the PWM control mode with the ZCD function is according to voltage signal V_{ZCD}. Unlike the PWM control mode, the PFM control mode does not generate the same number with a fixed switching frequency control, so the PFM control mode with the ZCD function can be useful in reducing switching and conduction losses at light loads.

To further understand the advantages and superiority of the PFM control mode with the ZCD function for buck converter to increase light load efficiency.

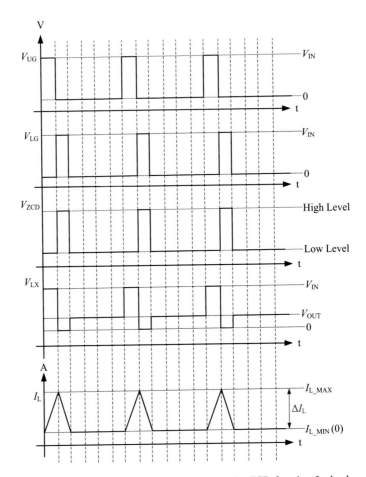

Fig. 1.31 Control signals of the PFM control mode with the ZCD function for buck converter

Experimental verifications were conducted to compare the feasibility and performance between the PFM control mode with the ZCD function and the PWM control mode for buck converter. To achieve a fair and correct comparison between the PFM control mode with the ZCD function and the PWM control mode for buck converter, so the experimental results are measured by the same operation conditions.

The specifications are as follows:

(1) Input voltage (V_{IN}): 12 V
(2) Output voltage (V_{OUT}): 1.05 V
(3) Output load current (I_{OUT}): 10 mA
(4) Switching frequency (F_S): 630 kHz

Fig. 1.32 Comparison of the experimental results between the PFM control mode with the ZCD function and the PWM control mode at the light load for buck converter

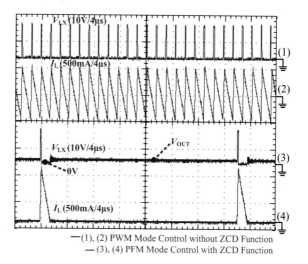

— (1), (2) PWM Mode Control without ZCD Function
— (3), (4) PFM Mode Control with ZCD Function

(5) MOSFET (S_A, S_B): BSC0909NS * 3
(6) Feedback resistors (R_{D1}, R_{D2}): 13.2 kΩ and 12 kΩ
(7) Main inductor (L): IHLP4040DZER1R0MA1 (1 μH)
(8) Output capacitors (C_O): 22 μF/6.3 V (R_{CO}: 3 mΩ)*2
(9) Reference voltage (V_{REF}): 0.5 V

Figure 1.32 compares the experimental results between the PFM control mode with the ZCD function and the PWM control mode at the light load for buck converter. Figure 1.32 represents the V_{LX} voltage and the I_L current the PFM control mode with the ZCD function and the PWM control mode at the light load. In Fig. 1.32, the red-colored waveforms represent the V_{LX} voltage and the I_L current for the PWM control mode. The blue-colored waveforms represent the V_{LX} voltage and the I_L current for the PFM control mode with the ZCD function. The signals are measured based on the same output load current. The switching frequency of the PWM control mode maintains 630 kHz at output load current 10 mA. The switching frequency of the PFM control mode with the ZCD function is based on the output load current condition and the switching frequency is measured 33.3 kHz at output load current 10 mA. The number of control pulses for the PWM control mode is larger than the PFM control mode with the ZCD function, so the PWM control mode has a large switching loss. In addition, the I_L current of the PFM control mode with the ZCD function is maintained zero current during 28 μs, so it is useful to save conduction loss on the S_B switch, because the inductor current is through the body diode of the S_B switch. Moreover, the ΔI_L ripple current is 1 A of the PFM control mode with the ZCD function and it is similar to the ΔI_L ripple current of the PWM control mode with the same width of the voltage signal V_{UG}. Based on these reasons, the PFM control mode with the ZCD function can significantly achieve a better efficiency at the light load.

References

1. P. Burger, Analysis of a class of pulse modulated dc-to-dc power converters, IEEE Trans. Ind. Elect. Contr. Instrum. **IECI-22**, 104 (1975)
2. L.E. Gallaher, Current regulator with AC and DC feedback, *U.S. Patent* 3 350 628, 31 Oct 1967
3. A.D. Schoenfeld, Y. Yu, ASDTIC control and standardized interface circuits applied to buck, parallel and buck-boost DC-to-DC power converters (*NASA, Washington, DC, NASA Rep. NASA* CR-121106, Feb. 1973)
4. C.W. Deisch, Switching control method changes power converter into a current source, in *Proceedings IEEE Power Electronics Specialists Conference*, (1978), pp. 300–306
5. P.L. Hunter, Converter circuit and method having fast responding current balance and limiting, U.S. Patent 4 002 963, 1 Nov 1977
6. L.H. Dixon, Average current-mode control of switching power supplies, in *Proceedings Unitrode Power Supply Design Seminars Handbook*, (1990), pp. 5.1–5.14
7. N. Mohan, Power electronics circuits: An overview," in *Proceedings IEEE Industrial Electronics Society Conference*, (1988), pp. 522–527
8. N. Mohan, W.P. Robbins, P. Imbertson, T.M. Undeland, R.C. Panaitescu, A.K. Jain, P. Jose, T. Begalke, Restructuring of first courses in power electronics and electric drives that integrates digital control. IEEE Trans. Power Electron. **18**, 429–437 (2003)
9. R.D. Middlebrook, S. Cuk, A general unified approach to modeling switching-converter power states, in *Proceedings of IEEE Power Electronics Specialists Conference,* (1976), pp. 18
10. D.Y. Chen, H.A. Owen, T.G. Wilson, Computer aided design and graphics applied to the study of inductor-energy-storage dc-to-dc electronic power converters. IEEE Trans. Aerosp. Electron. Syst. **AES-9**, 585 (1973)
11. R.W. Erickson, D. Maksimovic, *Fundamentals of power electronics* (Norwell, MA, Kluwer, 2001)
12. K.H. Chen, H.W. Huang, S.Y. Kuo, Fast-transient DC–DC converter with on-chip compensated error amplifier. IEEE Trans. Circuits Syst. II Express Briefs **54**(12), 1150–1154 (2007)
13. C.L. Chiu, K.H. Chen, A high accuracy current-balanced control technique for LED backlight, in *Proceendings of IEEE Power Electronics Specialists Conference*, (2008), pp. 4202–4206
14. D. Rand, B. Lehman, A. Shteynberg, Issues, models and solutions for triac modulated phase dimming of led lamps, in *2007 IEEE Power Electronics Specialists Conference*, (June 2007), pp. 1398–1404
15. A. Morrison, J.W. Zapata, S. Kouro, M.A. Perez, T.A. Meynard, H. Renaudineau, Partiat power dc-dc converter for photovoltaic two-stage string inverters, in *2016 IEEE Energy Conversion Congress and Exposition* (ECCE), (September 2016)
16. J.W. Zapata, T.A. Meynard, S. Kouro, Partial power dc-dc converter for large-scale photovoltaic systems, in *2016 IEEE 2nd Annual Southern Power Electronics Conference* (SPEC), (December 2016), pp. 1–6
17. S. Saponara, G. Pasetti, N. Costantino, F. Tinfena, P.D'Abramo, L. Fanucci, A flexible LED driver for automotive lighting applications: IC design and experimental characterization. IEEE Trans. Power Electron. **27**(3), 1071–1075 (2012)
18. C.-S. Moo, Y.-J. Chen, W.-C. Yang, An efficient driver for dimmable LED lighting. IEEE Trans. Power Electron. **27**(11), 4613–4618 (2012)
19. G. Sauerlander, D. Hente, H. Radermacher, E. Waffenschmidt, J. Jacobs, Driver electronics for LEDs, Proc. Conf. Rec. IEEE Ind. Appl. Conf. 2621–2626 (2006)
20. G. Rincon-Mora, P. Allen, A low-voltage low quiescent current low drop-out regulator. IEEE J. Solid-State Circuits. **33**(1), 36–44 (1998)

21. R.J. Milliken, J. Silva-Martínez, E. Sánchez-Sinencio, Full On-Chip CMOS Low-Dropout Voltage Regulator. IEEE Trans. Circuits Syst. I. Reg. Pap. **54**(9), 1879 (2007)
22. C. Chen, J. Wu, Z. Wang, 150 mA LDO with self-adjusting frequency compensation scheme, Electron. Lett. **47**(13), 767–768 (2011)
23. A. Patel, G. Rincon-Mora, High Power-Supply-Rejection (PSR) Current-Mode Low-Dropout (LDO) Regulator. IEEE Trans. Circuits Syst. II Exp. Briefs. **57**(11), 868–873 (2010)
24. E.N.Y. Ho, P.K.T. Mok, A capacitor-less CMOS active feedback low-dropout regulator with slew-rate enhancement for portable on-chip application. IEEE Trans. Circuits Syst. II Exp. Briefs. **57**(2), 80-84 (2010)
25. K.N. Leung, P. Mok, A capacitor-free CMOS low-dropout regulator with damping-factor-control frequency compensation. IEEE J. Solid-State Circuits. **38**(10), 1691–1702 (2003)
26. H. Lee, P. Mok, K.N. Leung, Design of low-power analog drivers based on slew-rate enhancement circuits for cmos low-dropout regulators. IEEE Trans. Circuits Syst. II Exp. Briefs. **52**(9), 563–567 (2005)
27. J.R. Huang, C.H. Wang, C.J. Lee, K.L. Tseng, D. Chen, Native AVP control method for constant output impedance of DC power converters, in *Proceedings of IEEE Power Electronics Specialists Conference*, (2007), pp. 2023–2028
28. A. Waizman, C.Y. Chung, Resonant free power network design using extended adaptive voltage positioning (EAVP) methodology. IEEE Trans. Adv. Packag. **24**, 236–244 (2001)
29. M. Lee, D. Chen, K. Huang, E. Tseng, B. Tai, Compensator design for adaptive voltage position (AVP) for multiphase VRMs, in *Proceedings of IEEE Power Electronics Specialists Conference*, (2006)
30. K. Yao, Y. Meng, P. Xu, F.C. Lee, Design considerations for VRM transient response based on the output impedance, in *Proceedings of IEEE Applied Power Electronics Conference and Exposition conference*, (2002), pp. 14–20
31. P.L. Wong, Performance improvements of multi-channel interleaving voltage regulator modules with integrated coupling inductors, Dissertation, Virginia Polytechnic Institute and State University, 2001
32. S.K. Mishra, Design-oriented analysis of modern active droop controlled power supplies. IEEE Trans. Ind. Electron. **56**(9), 3704–3708 (2009)
33. J.A.A. Qahouq, V. Arikatla, Power converter with digital sensorless adaptive voltage positioning control scheme. IEEE Trans. Ind. Electron. **58**(9), 4105–4116 (2010)
34. W.W. Chen, J.F. Chen, T.J. Liang, Dynamic Ramp control in current-mode adaptive on-time control for Buck converter on chip, in *Proceedings of IEEE Future Energy Electronics Conference and ECCE Asia (IFEEC 2017 - ECCE Asia)*, (2017), pp. 280–285
35. W.W. Chen, J.F. Chen, T.J. Liang, J.R. Huang, W.Y. Ting, Improved transient response using HFFC in current-mode CFCOT control for buck converter, in *Proceedings IEEE International Conference on Power Electronics and Drive Systems (PEDS)*, (2013), pp. 546–549
36. W.W. Chen, J.F. Chen, T.J. Liang, J.R. Huang, L.C. Wei, W.Y. Ting, Implementing dynamic quick response with high-frequency feedback control of the deformable constant on-time control for Buck converter on-chip. IET Power Electron. **6**(4), 383–391 (2013)
37. Richtek Tech. Corp., Analysis of Buck Converter Efficiency, Application Note, (2014)
38. J. Gallaghe, Coupled inductors improve multiphase buck efficiency. Power Electron. Technol. (2006)
39. Fairchild Semiconductor Inc., AN-6005 Synchronous buck MOSFET loss calculations with Excel model. Application Note, 2014
40. K. Yao, Y. Meng, F.C. Lee, A novel winding coupled-buck converter for high-frequency high step-down DC/DC conversion, in *Proceedings IEEE Power Electronics Specialists Conference*, (2002)

41. H. Krishnamurthy, V. Vaidya, P. Kumar, G. Matthew, S. Weng, B. Thiruvengadam, W. Proefrock, K. Ravichandran, V. De, A 500 MHz 68% efficient fully on-die digitally controlled buck voltage regulator on 22 nm tri-gate CMOS, in Symposium *VLSI Circuits Digest Technical Papers*, (2014), pp. 210–211

42. S.J. Kim, Q. Khan, M. Talegaonkar, A. Elshazly, A. Rao, N. Griesert, G. Winter, W. McIntyre, P. Hanumolu, High frequency buck converter design using time-based control techniques. IEEE J. Solid-State Circuits. **50**(4), 990–1001 (2015)

43. W.R. Liou, M.L. Yeh, Y.L. Kuo, A high efficiency dual-mode buck converter IC for portable applications, IEEE Trans. Power Electron. **23**(2), 667–677 (2008)

44. Texas Instruments Inc., MOSFET power losses and how they affect power-supply efficiency, *Application Report*, 2016

45. Texas Instruments Inc., Calculating Efficiency, *Application Report*, 2010

46. Texas Instruments Inc., Optimizing MOSFET characteristics by adjusting Gate Drive Amplitude, *Application Report*, 2005

47. A.V. Petershevs, S.R. Sanders, Digital multimode buck converter control with loss-minimizing synchronous rectifier adaptation. IEEE Trans. Power Electron. **21**(6), 1588–1599 (2006)

48. S. Angkititrakul, H. Hu, Design and analysis of buck converter with pulse-skipping modulation, in *Proceedings of IEEE Power Electronics Specialists Conference*, (2008), 1151–1156

49. X. Zhou, M. Donati, L. Amoroso, F.C. Lee, Improved light-load efficiency for synchronous rectifier voltage regulator module. IEEE Trans. Power Electron. **15**(5), pp. 826–834 (2000)

50. C.L. Chen, W.L. Hsieh, W.J. Lai, K.H. Chen, C.S. Wang, A new PWM/PFM control technique for improving efficiency over wide load range, *in Proceedings of IEEE International Conference on Electronics, Circuits and Systems*, (2008), pp. 962–965

51. S. Kapat, S. Banerjee, A. Patra, Discontinuous map analysis of a DC-DC converter governed by pulse skipping modulation. IEEE Trans. Circuits System. Part I. **57**(7), 1793–1801 (2010)

Chapter 2
Review of the PWM Control Circuits for Power Converters

2.1 Voltage-Mode Control Circuit for Power Converters

Power converters are electrical control circuits that transfer energy from a DC voltage source to the output loading and regulate the output voltage to meet the user design. Energy is transferred via electronic switches made with transistors and diodes to an output filter and then transferred to the output loading.

These converters employ square-wave PWM to achieve voltage regulation. The output voltage is regulated by varying the duty cycle of the power semiconductor switch driving signal. The voltage waveform across the switch and at the input of the filter is a square wave in nature and generally results in high switching losses when the switching frequency is increased. However, these converters are easy to control, are well understood, and have a wide load control range. These converters also operate at a fixed-frequency, variable duty cycle. This type of control signal is called PWM control signal. Depending on the duty cycle, these converters can operate in either CCM or discontinuous conduction mode (DCM). If the current through the output inductor never reaches zero, then the converter operates in CCM. If the current through the output inductor reaches zero, then the converter operates in DCM.

The output voltage is equal to the average value in the switching cycle of the voltage applied at the output filter. For real switches with parasitic elements, efficiency depends on conduction and switching losses, but the efficiency of power converters remains higher than that of linear regulators such as LDO.

Power converters are widely applied in portable electronic equipment and products, especially those designed to reduce standby power loss. They demonstrate high efficiency and present a fast transient response due to system design. The fixed-frequency PWM control scheme for power converters [1–24] is commonly used with current-mode and voltage-mode control circuits.

Voltage-mode control circuit is the simplest circuit structure for PWM control scheme for power converters. The major characteristic of this design is the presence

© Springer Nature Singapore Pte Ltd. 2018
W.-W. Chen and J.-F. Chen, *Control Techniques for Power
Converters with Integrated Circuit*, Power Systems,
https://doi.org/10.1007/978-981-10-7004-4_2

Fig. 2.1 Circuit diagram of
the voltage-mode control
circuit for buck converter

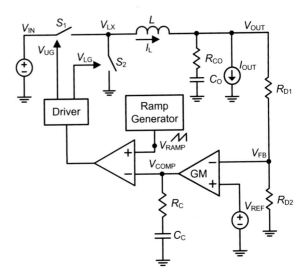

of a single voltage feedback path, with PWM performed by comparing the voltage error signal with a constant ramp waveform. Current limiting must be conducted separately. The advantages of voltage-mode control are as follows: the single feedback loop is easy to design and analyze and a large-amplitude ramp waveform provides good noise immunity for a stable modulation process. Figure 2.1 shows the circuit diagram of the voltage-mode control circuit for buck converter. S_1 and S_2 are the power switches integrated on-chip, L is the output inductor. R_{CO} is the ESR of output capacitor C_O. Current source I_{OUT} is the output load current. The driver circuit uses the input signal on-time width to generate two control signals V_{UG} and V_{LG}, these two signals should be avoid to turn on at the same time, because this operation make this system to have a shoot through problem. The compensator of R_C and C_C should be designed an optimization to increase the transient response. Only the feedback signal V_{FB} and reference voltage V_{REF} are built inside the IC. The output signal of comparator depends on the input signals V_{COMP} and V_{RAMP} results.

The PWM three-terminal model [23, 24] is a good tool to analyze loop stability because power switches S_1 and S_2 can be modeled as equivalent circuits similar to a dependent voltage source, a dependent current source, and an ideal transformer. The PWM three-terminal model with voltage-mode control circuit for the buck converter is shown in Fig. 2.2. In Fig. 2.2, a transfer function for control-to-output $G_{CTO}(s)$ is obtained with Eqs. (2.1)–(2.7). Based on the $G_{CTO}(s)$ transfer function, the voltage-mode control circuit for the buck converter has one zero and two poles system, as shown in Fig. 2.3. The DC gain of control-to-output depends on the input voltage by Eq. (2.2), and a large input voltage has a large DC gain. The voltage-mode control circuit for the buck converter is unsuitable for a wide input voltage range. The zero S_{Z1} depends on output capacitor C_O and ESR R_{CO} of the output capacitor. With regard to output capacitor selection, tantalum capacitors are

Fig. 2.2 PWM
three-terminal model with
voltage-mode control circuit
for buck converter

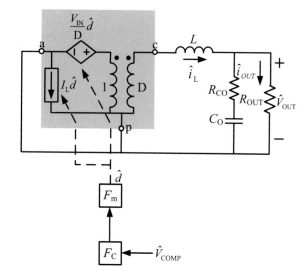

Fig. 2.3 Control-to-output
with one zero and two poles
system of voltage-mode
control circuit for buck
converter

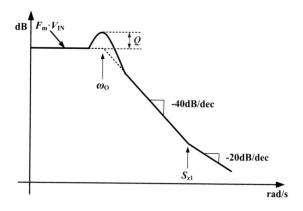

a poor choice for output capacitors because tantalum capacitors usually fail to create a short circuit across their terminals, thereby raising the possibility of a fire hazard. Ceramic or aluminum electrolytic capacitors are preferred because they do not have this failure mode. Ceramic capacitors possessing small footprint, low profile, low ESR, low cost, and high reliability are widely used in microprocessor decoupling and power converter filtering applications. Thus, ceramic capacitors are a good choice when the evaluation board area, cost reduction, or component height is considered. The low ESR of ceramic capacitors results in high performance and efficiency but may cause the system loop to become unstable. Generally, the ESR of a ceramic output is less than 10 mΩ, and the S_{Z1} zero location is set at a high frequency of the voltage-mode control circuit for the buck converter. Thus, the control-to-output $G_{CTO}(s)$ is changed to a two poles system.

$$G_{CTO}(s) = \frac{\hat{V}_{OUT}}{\hat{V}_{COMP}} = F_m \cdot V_{IN} \frac{1 + s/S_{Z1}}{1 + s/(\omega_O \cdot Q) + s^2/\omega_O^2} \tag{2.1}$$

$$G_{CTO}(0) = F_m \cdot V_{IN} \tag{2.2}$$

$$\omega_O = \frac{1}{\sqrt{L \cdot C_O}} \tag{2.3}$$

$$Q = R_{OUT} \cdot \sqrt{\frac{C_O}{L}} \tag{2.4}$$

$$F_m = \frac{1}{V_{RAMP}} \tag{2.5}$$

$$F_C = 1 \tag{2.6}$$

$$S_{Z1} = \frac{1}{C_O \cdot R_{CO}} \tag{2.7}$$

Table 2.1 lists the operation conditions of the voltage-mode control circuit for the buck converter. According to Table 2.1, the comparison of MathCAD predictions and SIMPLIS simulation results of the open-loop control-to-output bode plot for the buck converter is plotted in Fig. 2.4. In Fig. 2.4, the MathCAD predictions verify the SIMPLIS simulation results. The red-colored line represents the MathCAD predictions, and the blue-colored dot represents the SIMPLIS simulation results. The DC gain curve of the open-loop control-to-output bode plot is equal to 6.02 dB as obtained by Eq. (2.2). The two poles are located at 2.475 kHz by Eq. (2.3). The two poles cause a sharp phase drop of 180°, so the voltage-mode control circuit requires the addition of one zero to cancel the effect of the two poles. Based on these operation conditions using a large output capacitor, the zero is located at 18.09 kHz, so it can help increase the phase degree (30°). In general, the output capacitor uses a ceramic capacitor with high switching frequency, and the capacitance of a single capacitor should be less than 50 μF. Thus, the zero is mainly located at more than 300 kHz with 10 mΩ of ESR. The zero at high frequency cannot help increase the phase degree.

A buck converter can achieve step-down voltage from its input power supply to its output terminal, so it is also widely used to convert a computer's main supply

Table 2.1 Operation conditions in voltage-mode control circuit for buck converter

V_{IN} (V)	4.8 V	F_S (kHz)	500 kHz	L (μH)	4.7 μH
V_{OUT} (V)	1.2 V	R_{D1} (kΩ)	100 kΩ	C_O (μF)	880 μF
I_{OUT} (A)	5 A	R_{D2} (kΩ)	40 kΩ	R_{CO} (mΩ)	10 mΩ
V_{REF} (V)	0.8 V	V_{RAMP} (V)	2.4 V		

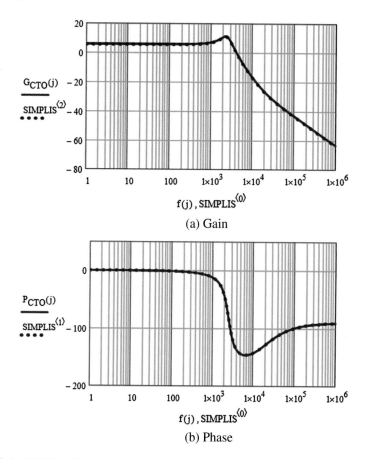

(a) Gain

(b) Phase

Fig. 2.4 MathCAD predictions of open-loop control-to-output bode plot for buck converter

voltage (often 12 V) down to lower voltages needed by USB, DRAM, and the CPU (1.8 V or less), or the output voltage is designed to be lower than the input power supply. A boost converter is different from a buck converter because a boost converter achieves step-up voltage from its input power supply to its output terminal; hence, it is widely used to convert a power supply voltage up to larger voltages needed by the display power driver and LED driver, or the output voltage is designed to be larger than input power supply.

Figure 2.5 shows a circuit diagram of the voltage-mode control circuit for boost converter. S_1 and S_2 are the power switches integrated on-chip, L is the output inductor. R_{CO} is the ESR of output capacitor C_O. Current source I_{OUT} is the output load current. The driver circuit uses the input signal on-time width to generate two control signals V_{UG} and V_{LG}, these two signals should be avoid to turn on at the same time, because this operation make this system to have a shoot through problem. The compensator of R_C and C_C should be designed an optimization to

Fig. 2.5 Circuit diagram of
the voltage-mode control
circuit for boost converter

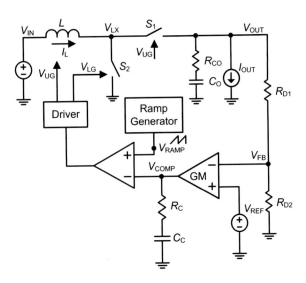

increase the transient response. Only the feedback signal V_{FB} and reference voltage
V_{REF} are built inside the IC. The output signal of comparator depends on the input
signals V_{COMP} and V_{RAMP} results.

The PWM three-terminal model [23, 24] with the voltage-mode control circuit
for the boost converter is shown in Fig. 2.6. In Fig. 2.6, a transfer function for

Fig. 2.6 PWM
three-terminal model with
voltage-mode control circuit
for boost converter

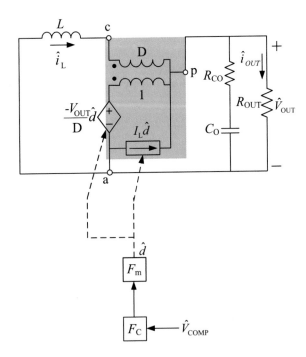

Fig. 2.7 Control-to-output
with one zero, one RHP zero,
and two poles system of
voltage-mode control circuit
for boost converter

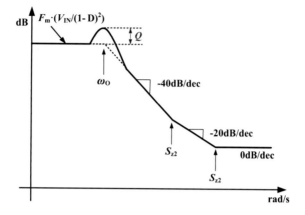

control-to-output $G_{CTO}(s)$ can be obtained by Eqs. (2.8)–(2.16). Based on the $G_{CTO}(s)$ transfer function, the voltage-mode control circuit for the boost converter has one zero, one right-half-plane (RHP) zero, and two poles system as shown in Fig. 2.7. The DC gain of control-to-output depends not only on the input voltage but also on the duty cycle according to Eq. (2.9). A large duty cycle has a large DC gain. Given this effect, the voltage-mode control circuit for the boost converter is unsuitable for generating very large output voltages. For example, if the duty cycle is 0.9, the DC gain of control-to-output should be increased to 40 dB with the same input voltage. A large DC gain of control-to-output makes it difficult to design an optimal compensator to ensure loop system stability. The first zero S_{Z1} depends on output capacitor C_O and ESR R_{CO} of the output capacitor. Ceramic capacitors possessing a small footprint, low profile, no failure mode, low ESR, low cost, and high reliability are widely used in microprocessor decoupling and power converter filtering applications. The low ESR of ceramic capacitors for output capacitors results in high performance and efficiency but may cause the system loop to become unstable. In general, the ESR of the ceramic output is less than 10 mΩ, and the first zero S_{Z1} location is set at a high frequency of the voltage-mode control circuit for the boost converter. Thus, the control-to-output $G_{CTO}(s)$ is changed to two poles and one RHP zero system. The second zero S_{Z2} is an RHP zero. RHP zero has the same 20 dB/decade rising gain magnitude as a conventional zero, but with a 90° phase lag instead of lead. This characteristic is difficult, if not impossible, to compensate. The designer is usually forced to roll off the loop gain at a relatively low frequency. The crossover frequency may be a decade or more below what it otherwise could be, resulting in severe impairment of the dynamic response. In general, RHP zero is set at a frequency higher than two poles for the boost converter. The RHP zero phase drop starts a decade earlier and negatively affects the potential phase margin of the converter's control loop. This is the nature of instability of a voltage-mode controlled boost converter running in CCM.

$$G_{CTO}(s) = \frac{\hat{V}_{OUT}}{\hat{V}_{COMP}} = \frac{V_{IN}}{(1-D)^2} \cdot F_m \cdot \frac{(1+s/S_{Z1}) \cdot (1-s/S_{Z2})}{1+s/(\omega_O \cdot Q) + s^2/\omega_O^2} \quad (2.8)$$

$$G_{CTO}(0) = \frac{V_{IN}}{(1-D)^2} \cdot F_m \quad (2.9)$$

$$\omega_O = \frac{1}{\sqrt{L_D \cdot C_O}} \quad (2.10)$$

$$Q = R_{OUT} \cdot \sqrt{\frac{C_O}{L_D}} \quad (2.11)$$

$$F_m = \frac{1}{V_{RAMP}} \quad (2.12)$$

$$F_C = 1 \quad (2.13)$$

$$S_{Z1} = \frac{1}{C_O \cdot R_{CO}} \quad (2.14)$$

$$S_{Z2} = \frac{R_{OUT}}{L_D} \quad (2.15)$$

$$L_D = \frac{L}{(1-D)^2} \quad (2.16)$$

Table 2.2 lists the operation conditions of the voltage-mode control circuit for the boost converter. According to Table 2.2, the comparison of MathCAD predictions and SIMPLIS simulation results of the open-loop control-to-output bode plot for the boost converter is plotted in Fig. 2.8. In Fig. 2.8, the MathCAD predictions verify the SIMPLIS simulation results. The red-colored line represents the MathCAD predictions, and the blue-colored dot represents the SIMPLIS simulation results. The DC gain curve of the open-loop control-to-output bode plot is equal to 25.666 dB according to Eq. (2.9). The two poles are located at 6.741 kHz by Eq. (2.10). The two poles cause a sharp phase drop of 180°, so the voltage-mode control circuit requires the addition of one zero to cancel the effect of the two poles. Based on this operation condition using the 44 μF output capacitor, the first zero S_{Z1} is located at 361.7 kHz by Eq. (2.14), and the second zero S_{Z2} is located at

Table 2.2 Operation conditions in voltage-mode control circuit for boost converter

V_{IN} (V)	5 V	F_S (kHz)	1000 kHz	L (μH)	2.2 μH
V_{OUT} (V)	12 V	R_{D1} (kΩ)	120 kΩ	C_O (μF)	44 μF
I_{OUT} (A)	0.3 A	R_{D2} (kΩ)	8.57 kΩ	R_{CO} (mΩ)	10 mΩ
V_{REF} (V)	0.8 V	V_{RAMP} (V)	1.5 V		

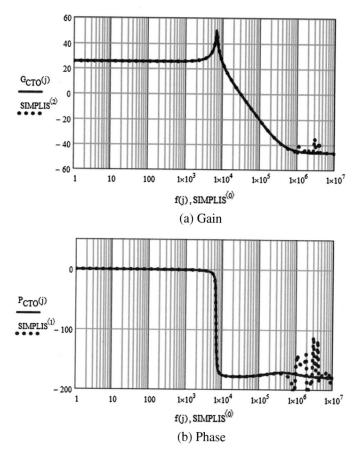

(a) Gain

(b) Phase

Fig. 2.8 Comparison of MathCAD predictions and SIMPLIS simulation results open-loop control-to-output bode plot for boost converter

502 kHz by Eq. (2.15). Thus, the first zero S_{Z1} can cancel the effect of the second zero S_{Z2}. The phase of open-loop control-to-output maintains a two-pole behavior similar to a sharp phase drop of 180°. Meanwhile, the RHP zero S_{Z2} depends on the output loading and duty cycle, so the user needs to consider an optimal compensator at the worst conditions to ensure that the RHP zero S_{Z2} does not affect the system loop stability.

2.2 Current-Mode Control Circuit for Power Converters

The current-mode control circuit contains two feedback control signals and differs from the voltage-mode control circuit. The output voltage is fed to an error amplifier to generate control signal V_{COMP}. Inductor current I_L is sampled into voltage signal

Fig. 2.9 Circuit diagram of
the current-mode control
circuit for buck converter

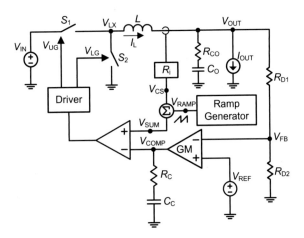

V_{CS}, and voltage signal V_{RAMP} is added for comparison with control signal V_{COMP} to control the peak inductor current and regulate output voltage V_{OUT} to the specified level. The inductor current I_L feedback control signal causes the current-mode control circuit to have a better transient response than the voltage-mode control circuit. The current-mode control circuit shows additional advantages of improved closed-loop stability and fast transient response [2–5, 8–22].

Figure 2.9 shows a circuit diagram of the current-mode control circuit for the buck converter. Figure 2.9 shows the circuit diagram of the current-mode control circuit for buck converter. S_1 and S_2 are the power switches integrated on-chip, L is the output inductor. R_i is the current sense gain. R_{CO} is the ESR of output capacitor C_O. Current source I_{OUT} is the output load current. The driver circuit uses the input signal on-time width to generate two control signals V_{UG} and V_{LG}, these two signals should be avoid to turn on at the same time, because this operation make this system to have a shoot through problem. The compensator of R_C and C_C should be designed an optimization to increase the transient response. Only the feedback signal V_{FB} and reference voltage V_{REF} are built inside the IC. The rising slope of inductor current I_L is sensed and required in the feedback loop as it contains supply information for the current-mode control circuit. The ramp generator circuit that can generate voltage signal V_{RAMP} to be added to inductor current I_L is sampled. The sum voltage of voltage signal V_{RAMP} and voltage signal V_{CS} is used through a resistor to obtain voltage signal V_{SUM}. The output signal of the comparator depends on the results of input signals V_{COMP} and V_{SUM}.

However, the current-mode control circuit produces a subharmonic issue [2–5, 14–22, 25–27] when the duty cycle is more than 50%, as shown in Fig. 2.10. Figure 2.10a shows that when the duty cycle is more than 50%, the duty cycle cannot maintain the same width of control signal V_{UG}, and a subharmonic issue with control signal V_{UG} arises. Figure 2.10b shows that when the duty cycle is less than 50%, the duty cycle can maintain the same width of control signal V_{UG}, so no subharmonic issue arises with control signal V_{UG}. The subharmonic issue can

Fig. 2.10 Without the ramp compensation of current-mode control circuit for buck converter

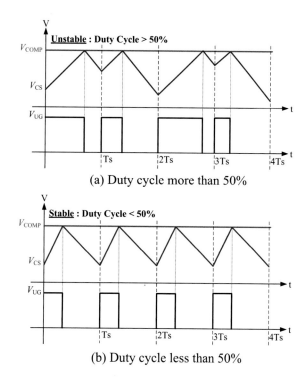

(a) Duty cycle more than 50%

(b) Duty cycle less than 50%

generate a different width of the duty cycle and may affect the output voltage ripple. An output voltage ripple with a subharmonic issue is larger than without a subharmonic issue.

To overcome the subharmonic issue, the ramp compensation signal is always added to the control loop to prevent subharmonic, as shown in Fig. 2.11. Ramp compensation is designed according to the inductor current ripple determined by the input voltage, output voltage, and inductance. A general power control IC often embeds a fixed ramp into the control loop. However, different operation conditions

Fig. 2.11 With the ramp compensation of current-mode control circuit for buck converter during duty cycle more than 50%

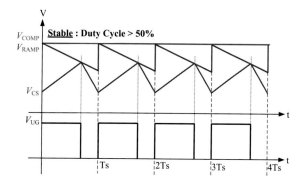

Fig. 2.12 PWM
three-terminal model with
current-mode control circuit
for buck converter

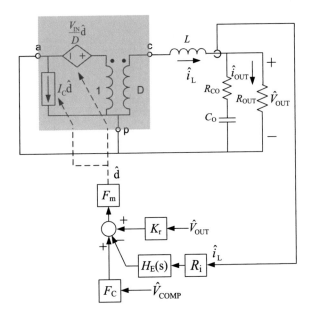

for input and output voltages can result in a varied inductor current ripple. Meeting
the fixed ramp compensation requires that the inductance be changed to maintain
the same current ripple. Thus, the compensation circuit should also be modified to
keep the same system loop stable.

The PWM three-terminal model [23, 24] with the current-mode control circuit
for the buck converter is shown in Fig. 2.12. In Fig. 2.12 a transfer function for
control-to-output $G_{CTO}(s)$ can be obtained by Eqs. (2.17)–(2.28). $F_P(s)$ provides
one pole and one zero [14–22]. The pole of $F_P(s)$ is located at the dominant
low-frequency characteristic of the system in Eqs. (2.19)–(2.20). $F_H(s)$ provides
two poles at half the switching frequency with good ramp compensation design in
Eqs. (2.22)–(2.24). The high-frequency effects are ignored. However, if the m_C
value is small, it presents no ramp compensation or small ramp compensation.
A small m_C value has two poles, which can cause a high Q at half of the switching
frequency. If the ramp compensation is large, it presents large ramp compensation.
At a large m_C value, given that the ramp compensation is much larger than the
inductor current, the control loop is formed through a voltage-mode control
behavior. The current-mode control is more stable than the voltage-mode control
with only one dominant pole.

$$G_{CTO}(s) = \frac{\hat{V}_{OUT}}{\hat{V}_{COMP}} = \frac{R_{OUT}}{R_i} \cdot \frac{1}{1 + \frac{R_{OUT}}{L \cdot F_S}\left[\left(1 + \frac{S_E}{S_N}\right) \cdot D' - 0.5\right]} \cdot Fp(s) \cdot Fh(s)$$

$$(2.17)$$

$$G_{CTO}(0) = \frac{R_{OUT}}{R_i} \cdot \frac{1}{1 + \frac{R_{OUT}}{L \cdot F_S} \left[\left(1 + \frac{S_E}{S_N} \right) \cdot D' - 0.5 \right]} \tag{2.18}$$

$$F_P(s) = \frac{1 + s/S_{Z1}}{1 + s/S_P} \tag{2.19}$$

$$S_P = \frac{1}{C_O \cdot R_{OUT}} + \frac{1}{C_O \cdot L \cdot F_S} \cdot \left[\left(1 + \frac{S_E}{S_N} \right) \cdot D' - 0.5 \right] \tag{2.20}$$

$$S_{Z1} = \frac{1}{C_O \cdot R_{CO}} \tag{2.21}$$

$$F_h(s) = \frac{1}{1 + \frac{s}{\omega_n \cdot Q_P} + \frac{s^2}{\omega_n^2}} \tag{2.22}$$

$$Q_P = \frac{1}{\pi \cdot \left[\left(1 + \frac{S_E}{S_N} \right) \cdot D' - 0.5 \right]} \tag{2.23}$$

$$\omega_n = \pi \cdot F_S \tag{2.24}$$

$$F_C = 1 \tag{2.25}$$

$$m_C = 1 + \frac{S_E}{S_N} \tag{2.26}$$

$$S_N = \frac{V_{IN} - V_{OUT}}{L} \cdot R_i \tag{2.27}$$

$$S_E = V_{RAMP} \cdot F_S \tag{2.28}$$

The DC gain of control-to-output directly depends on the output loading and current sense gain in Eq. (2.18). The DC gain of control-to-output of the current-mode control circuit is different from that of the voltage-mode control circuit because the former is not directly dependent on the input voltage. Thus, the current-mode control circuit is suitable for a wide range of input voltages. Moreover, the current-mode control circuit has current feedback loop control, so the DC gain is highly related to the current information on output loading and current sense gain.

The zero S_{Z1} depends on output capacitor C_O and the ESR R_{CO} of the output capacitor in Eq. (2.21). Low ESR of a ceramic capacitor for output capacitors is less than 10 mΩ, and the S_{Z1} zero location is set at a high frequency of the current-mode control circuit for the buck converter. The pole S_P of control-to-output is the dominant low frequency characteristic of the system in Eq. (2.20). It is different from the voltage-mode control circuit because the inductor current I_L feedback

Table 2.3 Operation conditions in current-mode control circuit for buck converter

V_{IN} (V)	5 V	F_S (kHz)	380 kHz	L (µH)	15 µH
V_{OUT} (V)	3.3 V	R_{D1} (kΩ)	125 kΩ	C_O (µF)	22 µF
I_{OUT} (A)	1 A	R_{D2} (kΩ)	40 kΩ	R_{CO} (mΩ)	10 mΩ
V_{REF} (V)	0.8 V	R_i (A/V)	0.4 A/V	V_{RAMP} (V)	0.358 mV

control signal can be similar to a zero behavior, so the current-mode control circuit is approximately one pole less than the voltage-mode control circuit.

Table 2.3 lists the operation conditions of the current-mode control circuit for the buck converter. Based on Table 2.3, a comparison of MathCAD predictions and SIMPLIS simulation results of the open-loop control-to-output bode plot for the buck converter is provided in Fig. 2.13. In Fig. 2.13, the MathCAD predictions verify the SIMPLIS simulation results. The red-colored line represents the

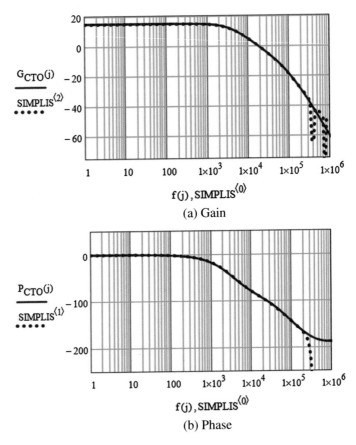

(a) Gain

(b) Phase

Fig. 2.13 MathCAD predictions of open-loop control-to-output bode plot for buck converter

Fig. 2.14 Circuit diagram of
the current-mode control
circuit for boost converter

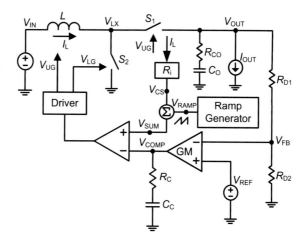

MathCAD predictions, and the blue-colored dot represents the SIMPLIS simulation results. The DC gain curve of the open-loop control-to-output bode plot is equal to 14.819 dB by Eq. (2.18). The pole S_P of control-to-output is located at 3.284 kHz by Eq. (2.20). Based on this operation condition using the small output capacitor, the zero is located at 723.4 kHz, so it cannot help increase the phase degree. However, two poles of $F_H(s)$ exert an effect on the phase degree in these operation conditions, and the phase degree is $-140°$ at 100 kHz.

Figure 2.14 shows a circuit diagram of the current-mode control circuit for the boost converter. S_1 and S_2 are the power switches integrated on-chip, L is the output inductor. R_i is the current sense gain. R_{CO} is the ESR of output capacitor C_O. Current source I_{OUT} is the output load current. The driver circuit uses the input signal on-time width to generate two control signals V_{UG} and V_{LG}, these two signals should be avoid to turn on at the same time, because this operation make this system to have a shoot through problem. The compensator of R_C and C_C should be designed an optimization to increase the transient response. Only the feedback signal V_{FB} and reference voltage V_{REF} are built inside the IC. The rising slope of inductor current I_L is sensed and required in the feedback loop as it contains supply information for the current-mode control circuit. The ramp generator circuit that can generate voltage signal V_{RAMP} to be added to inductor current I_L is sampled. The sum voltage of voltage signal V_{RAMP} and voltage signal V_{CS} is used through a resistor to obtain voltage signal V_{SUM}. The output signal of the comparator depends on the results of input signals V_{COMP} and V_{SUM}.

The PWM three-terminal model [23, 24] with the current-mode control circuit for the boost converter is shown in Fig. 2.15. In Fig. 2.15, a transfer function for control-to-output $G_{CTO}(s)$ can be obtained by Eqs. (2.29)–(2.45). $F_P(s)$ provides one pole, one zero, and one RHP zero by Eqs. (2.34)–(2.38) [14–22]. The pole of $F_P(s)$ is located at the dominant low-frequency characteristic of the system by Eqs. (2.34)–(2.35). $F_H(s)$ provides two poles at half the switching frequency with a good ramp compensation design by Eqs. (2.39)–(2.41). The high-frequency effects

Fig. 2.15 PWM
three-terminal model with
current-mode control circuit
for boost converter

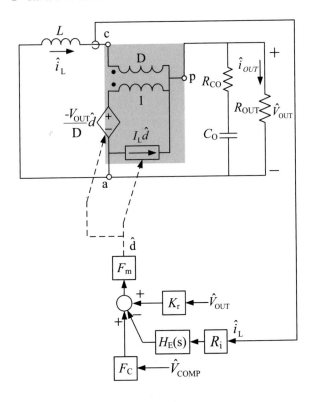

are ignored. However, if the m_C value is small, it presents no ramp compensation or small ramp compensation. The current-mode control is more stable than the voltage-mode control with only one dominant pole.

$$G_{CTO}(s) = \frac{\hat{V}_{OUT}}{\hat{V}_{COMP}} = \frac{(1-D) \cdot R_{OUT}}{KD \cdot R_i} \cdot Fp(s) \cdot Fh(s) \qquad (2.29)$$

$$G_{CTO}(0) = \frac{(1-D) \cdot R_{OUT}}{KD \cdot R_i} \qquad (2.30)$$

$$KD = 2 + \frac{(1-D)^2 \cdot R_{OUT}}{KD \cdot R_i}\left[\frac{1}{Km} + \frac{K}{(1-D)}\right] \qquad (2.31)$$

$$Km = \frac{1}{(0.5-D) \cdot R_i \cdot \frac{T_S}{L} + \frac{V_{RAMP}}{V_{OUT}}} \qquad (2.32)$$

$$K = 0.5 \cdot R_i \cdot \frac{T_S}{L} \cdot (1-D) \cdot D \qquad (2.33)$$

$$F_P(s) = \frac{(1 + s/S_{Z1}) \cdot (1 - s/S_{Z2})}{1 + s/S_P} \tag{2.34}$$

$$S_P = \frac{KD}{C_O \cdot R_{OUT}} \tag{2.35}$$

$$S_{Z1} = \frac{1}{C_O \cdot R_{CO}} \tag{2.36}$$

$$S_{Z2} = \frac{R_{OUT}}{L_D} \tag{2.37}$$

$$L_D = \frac{L}{(1 - D)^2} \tag{2.38}$$

$$F_h(s) = \frac{1}{1 + \frac{s}{\omega_n \cdot Q_P} + \frac{s^2}{\omega_n^2}} \tag{2.39}$$

$$Q_P = \frac{1}{\pi \cdot \left[\left(1 + \frac{S_E}{S_N}\right) \cdot D' - 0.5\right]} \tag{2.40}$$

$$\omega_n = \pi \cdot F_S \tag{2.41}$$

$$F_C = 1 \tag{2.42}$$

$$m_C = 1 + \frac{S_E}{S_N} \tag{2.43}$$

$$S_N = \frac{V_{IN}}{L} \cdot R_i \tag{2.44}$$

$$S_E = V_{RAMP} \cdot F_S \tag{2.45}$$

The first zero S_{Z1} depends on output capacitor C_O and ESR R_{CO} of the output capacitor. Low ESR of a ceramic capacitor for output capacitors is less than 10 mΩ, and the first zero S_{Z1} location is set at a high frequency of the current-mode control circuit for the boost converter. Thus, the control-to-output $G_{CTO}(s)$ is changed to one pole and one RHP zero system. The second zero S_{Z2} is an RHP zero. RHP zero has the same 20 dB/decade rising gain magnitude as a conventional zero but with a 90° phase lag instead of lead. This characteristic is difficult, if not impossible, to compensate. The designer is usually forced to roll off the loop gain at a relatively low frequency. The crossover frequency may be a decade or more below what it otherwise could be, resulting in severe impairment of the dynamic response. In general, RHP zero is at a higher frequency than the dominant pole for the boost

Table 2.4 Operation conditions in voltage-mode control circuit for boost converter

V_{IN} (V)	5 V	F_S (kHz)	1200 kHz	L (μH)	470 nH
V_{OUT} (V)	13.6 V	R_{D1} (kΩ)	26.9 kΩ	C_O (μF)	80 μF
I_{OUT} (A)	0.6 A	R_{D2} (kΩ)	2.7 kΩ	R_{CO} (mΩ)	2 mΩ
V_{REF} (V)	1.24 V	R_i (A/V)	0.1 A/V	V_{RAMP} (V)	0.872 mV

converter. The RHP zero phase drop starts a decade earlier and negatively affects the potential phase margin of the converter's control loop.

Table 2.4 lists the operation conditions of the current-mode control circuit for the boost converter. Based on Table 2.4, a comparison of MathCAD predictions and SIMPLIS simulation results of the open-loop control-to-output bode plot for the boost converter is shown in Fig. 2.16. In Fig. 2.16, the MathCAD predictions verify the SIMPLIS simulation results. The red-colored line represents the

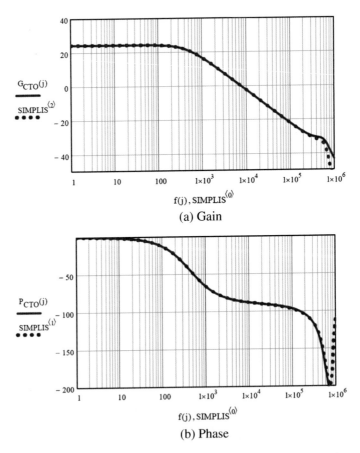

(a) Gain

(b) Phase

Fig. 2.16 Comparison of MathCAD predictions and SIMPLIS simulation results open-loop control-to-output bode plot for boost converter

MathCAD predictions, and the blue-colored dot represents the SIMPLIS simulation results. The DC gain curve of the open-loop control-to-output bode plot is equal to 24.386 dB by Eq. (2.30) using MathCAD, and the DC gain curves of the open-loop control-to-output bode plot using the SIMPLIS simulation tool is 23.38 dB. The MathCAD predictions verify the SIMPLIS simulation results. Pole S_P of control-to-output is located at 442.06 Hz by Eq. (2.32). Based on this operation condition using the 80 µF output capacitor, the first zero S_{Z1} is located at 994.7 kHz by Eq. (2.33), and the second zero S_{Z2} is located at 1010 kHz by Eq. (2.34). At this operation condition, the open-loop control-to-output bode plot shows one pole at the dominant low-frequency characteristics of the system only, so the current-mode control is more stable than the voltage-mode control with only one dominant pole.

RHP zero S_{Z2} depends on the output loading and duty cycle, so the user needs to consider an optimal compensator at the worst conditions to ensure that RHP zero S_{Z2} does not affect system loop stability.

2.3 Compensation Design for Power Converters

The voltage-mode and current-mode control circuits for power converters can obtain different control-to-output transfer functions. The control-to-output transfer function can indicate the number of poles and zeroes and their locations. In a power converter, the system control loop is a closed-loop loop gain and has a feedback loop to regulate the suitable duty cycle to control the power switches and obtain the required output voltage and output loading. In general, feedback loop consists of feedback resistors and a compensator. The purpose of the feedback resistors is to design the required output voltage. The required output voltage depends on reference voltage V_{REF} by Eq. (2.46).

$$V_{OUT} = V_{REF} \cdot \left(1 + \frac{R_{D1}}{R_{D2}}\right) \tag{2.46}$$

The purpose of adding a compensator to the error amplifier is to compensate for some of the gains and phases contained in the control-to-output transfer function and feedback resistors to achieve high stability of the power supply. The ultimate goal is to make the overall closed-loop loop gain transfer function satisfy the stability criteria. Figure 2.17 shows a block diagram of closed-loop loop gain for the current-mode control circuit. Closed-loop loop gain $T(s)$ contains the open-loop control-to-output transfer function $G_{CTO}(s)$, the feedback resistors' transfer function $G_{FB}(s)$, and the compensator's transfer function $G_{COMP}(s)$.

In general, open-loop control-to-output transfer function $G_{CTO}(s)$ and the feedback resistors' transfer function $G_{FB}(s)$ are fixed under operation conditions, so the user or designer design the optimal compensator transfer function $G_{COMP}(s)$ according to the open-loop control-to-output transfer function $G_{CTO}(s)$ and the feedback resistors' transfer function $G_{FB}(s)$ to satisfy the stability criteria in

Fig. 2.17 Block diagram of closed-loop loop gain for current-mode control circuit

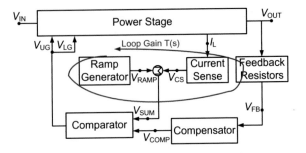

Eq. (2.47). Based on this equation, the information of gain margin (GM) and phase margin (PM) can be easily obtained to understand system loop stability.

$$1 + T(s) < 0 \tag{2.47}$$

This stability criterion of gain margin is the phase of closed-loop loop gain is close to $-180°$, and the desired gain margin anywhere the gain is smaller than 0 dB by Eq. (2.48). The gain margin is not defined and considered a safe value. In general, the gain margin is smaller than -10 dB at the least, as shown in Fig. 2.18. If the phase of closed-loop loop gain is not smaller than $-180°$, the gain margin would be difficult to find.

$$GM = 0\,dB - T(s)|_{\angle T(s) = -180°} \tag{2.48}$$

The stability criterion of the gain margin of closed-loop loop gain is close to 0 dB, and the desired phase margin anywhere the phase is greater than $-180°$ by

Fig. 2.18 Definition of Gain margin and phase margin for closed-loop loop gain

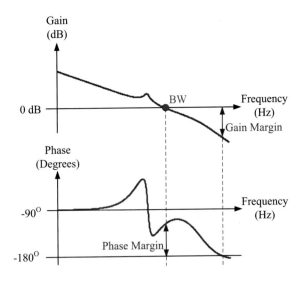

Eq. (2.49). The phase margin is generally considered a safe value when it is greater than 45° and yields a well-damped transient load response, as shown in Fig. 2.18.

$$\text{PM} = 180° \, 0\,\text{dB} + \angle T(s)\big|_{|T(s)|=0\,\text{dB}} \tag{2.49}$$

Having the slope of the gain curve at crossover frequency F_C with a value of −20 dB/decade is desirable because the phase variation is not a sharp phase drop of 180° at the crossover frequency, and it exerts an effect on reducing the phase margin. The slope of the gain curve is designed to be −40 dB/decade to exceed crossover frequency F_C and to increase noise immunity. The impact on the phase margin must be avoided because a good phase margin is prioritized over good noise immunity.

An optimal compensator can allow this system to achieve high system loop stability. It not only makes this system acquire good phase and gain margins but also increases the DC gain of the closed-loop loop gain. In general, control-to-output transfer function $G_{CTO}(s)$ and feedback resistors' transfer function $G_{FB}(s)$ possess a low DC gain. A low DC gain can cause steady-state errors, and large steady-state errors exhibit poor load regulation, resulting in load regulation results that are out of specifications for IC products. Moreover, large steady-state errors may regulate a failure output voltage, thus causing the system to operate under error conditions.

Figure 2.19 shows a schematic of the operational amplifier (OPA) type-II compensator for AC analysis. The operational amplifier is the basis of the closed-loop system and can be modeled similar to a voltage-controlled voltage source (VCVS). In a feedback system, OPA amplifies the error detected between feedback voltage V_{FB} and reference voltage V_{REF}. Under steady DC conditions, the input terminals of feedback voltage V_{FB} and reference voltage V_{REF} are virtually at the same voltage level. However, although both feedback resistors of the voltage

Fig. 2.19 Schematic of the OPA type-II compensator for AC analysis

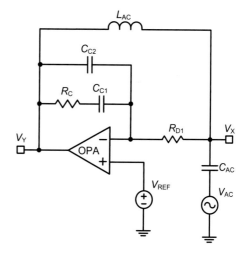

divider affect the DC level of the output voltage, from the AC point of view, only feedback resistor R_{D1} is used in the AC analysis. Feedback resistor R_{D2} is considered and designed for output voltage requirement. Therefore, feedback resistor R_{D2} is usually ignored in AC analysis. The OPA type-II compensator bode plot cannot be directly simulated by SIMPLIS or HSpice without initial conditions, so the simulation circuit usually uses inductor L_{AC} and capacitor C_{AC} to generate the initial conditions. Then, the system's bode plot can be simulated. Inductor L_{AC} is used to short output terminal V_Y to input terminal V_X at the DC level, and the initial conditions of voltage V_Y is equal to those of voltage V_X. Moreover, capacitor C_{AC} is used to block the AC signal injected into input terminal V_X. To avoid changes in the OPA type-II compensator bode plot, inductor L_{AC} should be designed with large inductance of 1 GH, and capacitor C_{AC} should be designed with large inductance of 1 GF.

In Fig. 2.19, a transfer function for OPA type-II compensator $G_{COMP}(s)$ can be obtained by Eqs. (2.50)–(2.53). Two poles and one zero are used for AC analysis, as shown in Fig. 2.20. One pole is an initial pole, and this pole crosses over 0 dB at S_C. Regarding the location of pole S_{P1} and zero S_{Z1}, zero S_{Z1} should be smaller than pole S_{P1}. The OPA type-II compensator may increase system loop stability, so capacitor C_{C1} should be much larger than capacitor C_{C2} to ensure optimal system loop stability. Moreover, pole S_{P1} is designed at a high frequency to decay the high-frequency noise, so pole S_{P1} can increase the system loop to achieve good noise immunity. In Fig. 2.20, the OPA type-II compensator has one zero to increase the phase degrees, and the maximum phase degrees reach 90°. Thus, the OPA type-II compensator is widely used as a voltage-mode control circuit for the buck converter, current-mode control circuit for the buck converter, or current-mode control circuit for the boost converter.

Fig. 2.20 Pole-Zero location of OPA type-II compensator for AC analysis

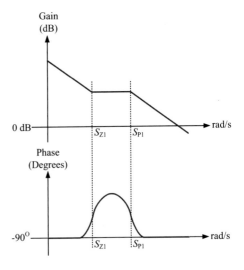

$$G_{\text{COMP}}(s) = \frac{\hat{V}_Y}{\hat{V}_X} = \frac{(1 + s/S_{Z1})}{s \cdot (C_{C1} \cdot R_{D1}) \cdot (1 + s/S_{P1})} \tag{2.50}$$

$$S_{P1} = \frac{1}{C_{C2} \cdot R_C} \tag{2.51}$$

$$S_{Z1} = \frac{1}{C_{C1} \cdot R_C} \tag{2.52}$$

$$S_C = \frac{1}{C_{C1} \cdot R_{D1}} \tag{2.53}$$

In power ICs, most products use an operational transconductance amplifier (OTA) instead of an operational amplifier because the output signal of OTA is a current signal and not a voltage signal. Hence, OTA does not need the circuit to convert a current signal to a voltage signal. OTA can be modeled similar to a VCCS. Figure 2.21 shows a schematic of OTA type-II compensator for AC analysis. In Fig. 2.21, resistor R_C and capacitors C_{C1} and C_{C2} are not connected between the input and output terminals. They are connected from the output terminal to the ground. Resistor R_O is an equivalent resistor at the output terminal and is a factor that affects the DC gain of the OTA type-II compensator. OTA is the basis of the closed-loop system. In a feedback system, OTA is used to amplify the error detected between feedback voltage V_{FB} and reference voltage V_{REF}. Under steady DC conditions, the input terminals of feedback voltage V_{FB} and reference voltage V_{REF} are virtually at the same voltage level. However, although both feedback resistors of the voltage divider affect the DC level of the output voltage, from the AC point of view, feedback resistors R_{D1} and R_{D2} do not affect the AC analysis. Therefore, feedback resistors R_{D1} and R_{D2} are usually ignored in AC analysis.

In general, the OTA type-II compensator bode plot cannot be directly simulated by SIMPLIS or HSpice without initial conditions. The simulation circuit usually uses inductor L_{AC} and capacitor C_{AC} to generate the initial conditions. Then, the bode plot of the system can be simulated. In Fig. 2.21, a transfer function for OTA

Fig. 2.21 Schematic of the OTA type-II compensator for AC analysis

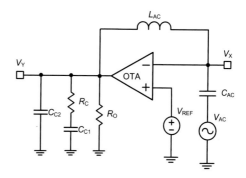

Fig. 2.22 Pole-Zero location
of OTA type-II compensator
for AC analysis

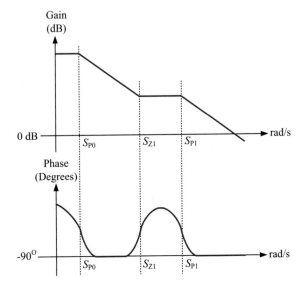

type-II compensator $G_{COMP}(s)$ can be obtained by Eqs. (2.54)–(2.58). Two poles
and one zero are used for AC analysis, as shown in Fig. 2.22. The DC gain of the
OTA type-II compensator depends on gain G_M of OTA and output equivalent
resistor R_O, so the DC gain of the OTA type-II compensator is different from the
DC gain of the OPA type-II compensator. Pole S_{P0} is an initial pole located at a low
frequency.

Regarding the location of pole S_{P1} and zero S_{Z1}, zero S_{Z1} should be designed
smaller than pole S_{P1}. The OTA type-II compensator may increase system loop
stability, so capacitor C_{C1} should be much larger than capacitor C_{C2} to ensure
optimal system loop stability. Pole S_{P1} is designed at a high frequency to decay the
high-frequency noise, so pole S_{P1} can increase the system loop to achieve good
noise immunity. In Fig. 2.22, the OTA type-II compensator has one zero to increase
the phase degrees, and the maximum phase degrees reach 90°. Thus, the OTA
type-II compensator is widely used as a voltage-mode control circuit for the buck
converter, current-mode control circuit for the buck converter, or current-mode
control circuit for the boost converter.

$$G_{COMP}(s) = \frac{\hat{V}_Y}{\hat{V}_X} = G_M \cdot R_O \frac{(1+s/S_{Z1})}{(1+s/S_{P0}) \cdot (1+s/S_{P1})} \qquad (2.54)$$

$$G_{COMP}(0) = \frac{\hat{V}_Y}{\hat{V}_X} = G_M \cdot R_O \qquad (2.55)$$

$$S_{P0} = \frac{1}{C_{C1} \cdot R_O} \qquad (2.56)$$

$$S_{P1} = \frac{1}{C_{C2} \cdot R_C} \tag{2.57}$$

$$S_{Z1} = \frac{1}{C_{C1} \cdot R_C} \tag{2.58}$$

The OPA type-II compensator and OTA type-II compensator for AC analysis have two poles and one zero structure. If the open-loop control-to-output has a two-pole system at the dominant frequency, the compensator should be implemented by the OPA type-III compensator or the OTA type-III compensator because these compensators have three poles and two zeros structure. The designer or user can design one pole at a high frequency or remove one pole to affect the closed-loop loop gain.

Figure 2.23 shows a schematic of the OPA type-III compensator for AC analysis. The operational amplifier is the basis of the closed-loop system. Feedback resistors R_{D1} and R_{D2} of the voltage divider affect the DC level of the output voltage, but from the AC point of view, only feedback resistor R_{D1} is used in the AC analysis. Feedback resistor R_{D2} is considered and designed for output voltage requirement. Therefore, feedback resistor R_{D2} is usually ignored in AC analysis. In general, the OPA type-III compensator bode plot cannot be directly simulated by SIMPLIS or HSpice without initial conditions, so the simulation circuit usually uses inductor L_{AC} and capacitor C_{AC} to generate the initial conditions. Then, the system's bode plot can be simulated.

In Fig. 2.23, a transfer function for OPA type-III compensator $G_{COMP}(s)$ can be obtained by Eqs. (2.59)–(2.64). Three poles and two zeros are used for AC analysis, as shown in Fig. 2.24. One pole is an initial pole, and this pole crosses over 0 dB at S_C. Regarding the location of pole S_{P1} and zero S_{Z1}, zero S_{Z1} should be designed smaller than pole S_{P1}. The OPA type-III compensator may increase the system loop stability, so capacitor C_{C1} should be much larger than capacitor C_{C2} to ensure

Fig. 2.23 Schematic of the OPA type-III compensator for AC analysis

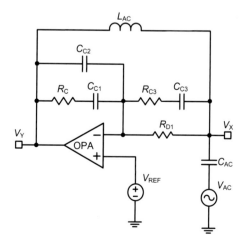

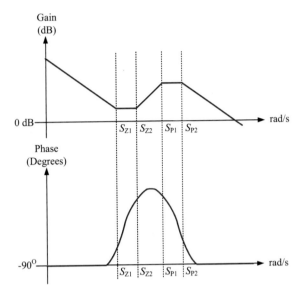

Fig. 2.24 Pole-Zero location of OPA type-III compensator for AC analysis

optimal system loop stability. Moreover, pole S_{P1} is designed at a high frequency to decay high-frequency noise, so pole S_{P1} can increase the system loop to achieve good noise immunity. In addition, regarding the location of pole S_{P2} and zero S_{Z2}, zero S_{Z2} must be formed smaller than pole S_{P2} by Eqs. (2.62)–(2.63) because the sum of two resistors R_{C3} and R_{D1} are larger than resistor R_{C3} only.

$$G_{COMP}(s) = \frac{\hat{V}_Y}{\hat{V}_X} = \frac{(1 + s/S_{Z1}) \cdot (1 + s/S_{Z2})}{s \cdot (C_{C1} \cdot R_{D1}) \cdot (1 + s/S_{P1}) \cdot (1 + s/S_{P2})} \tag{2.59}$$

$$S_{P1} = \frac{1}{C_{C2} \cdot R_C} \tag{2.60}$$

$$S_{Z1} = \frac{1}{C_{C1} \cdot R_C} \tag{2.61}$$

$$S_{P2} = \frac{1}{C_{C3} \cdot R_{C3}} \tag{2.62}$$

$$S_{Z2} = \frac{1}{C_{C3} \cdot (R_{C3} + R_{D1})} \tag{2.63}$$

$$S_C = \frac{1}{C_{C1} \cdot R_{D1}} \tag{2.64}$$

In Fig. 2.24, the OPA type-III compensator has two zeros to increase the phase degrees, and the maximum phase degrees reach 180°. Thus, the OPA type-III compensator is widely used as a voltage-mode control circuit for the buck converter or voltage-mode control circuit for the boost converter. If open-loop control-to-output has two poles at the dominant frequency, designers or users must apply the OPA type-III compensator, and the closed-loop loop gain may suffer from a bad phase margin with a large bandwidth or bad noise immunity. This implemented circuit also needs two passive components for R_{C3} and C_{C3}.

To increase the phase degrees in several real case designs for the OPA type-III compensator, the designer or user usually removes pole S_{P2} of transfer function $G_{COMP}(s)$ for the OPA type-III compensator. A good option is to design resistor R_{C3} for an mΩ level or remove resistor R_{C3} directly. Two poles and two zeros are used for AC analysis, as shown in Fig. 2.25. Transfer function $G_{COMP}(s)$ can be obtained and rewritten for the OPA type-III compensator by Eqs. (2.65)–(2.69).

$$G_{COMP}(s) = \frac{\hat{V}_Y}{\hat{V}_X} = \frac{(1 + s/S_{Z1}) \cdot (1 + s/S_{Z2})}{s \cdot (C_{C1} \cdot R_{D1}) \cdot (1 + s/S_{P1})} \qquad (2.65)$$

$$S_{P1} = \frac{1}{C_{C2} \cdot R_C} \qquad (2.66)$$

$$S_{Z1} = \frac{1}{C_{C1} \cdot R_C} \qquad (2.67)$$

Fig. 2.25 Pole-Zero location of OPA type-III compensator without the resistor R_{C3} for AC analysis

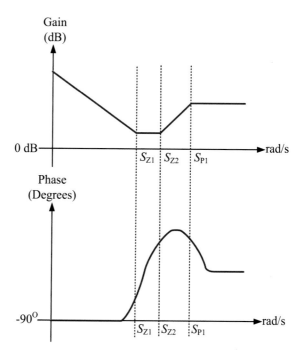

$$S_{Z2} = \frac{1}{C_{C3} \cdot R_{D1}} \tag{2.68}$$

$$S_C = \frac{1}{C_{C1} \cdot R_{D1}} \tag{2.69}$$

As shown in Fig. 2.25, the OPA type-III compensator has two zeros to increase the phase degrees, and the maximum phase degrees reach 180°. Thus, the OPA type-III compensator is widely used as a voltage-mode control circuit for the buck converter or voltage-mode control circuit for the boost converter. However, in the OPA type-III compensator without resistor R_{C3}, the closed-loop loop gain is needed to check the performance in phase margin and noise immunity.

Figure 2.26 shows a schematic of the OTA type-III compensator for AC analysis. The operational amplifier is the basis of the closed-loop system. Feedback resistors R_{D1} and R_{D2} of the voltage divider affect the DC level of the output voltage, but from the AC point of view, only feedback resistor R_{D1} is used in the AC analysis. It is different from the OTA type-II compensator for AC analysis. Therefore, feedback resistor R_{D1} cannot be ignored in AC analysis. Feedback resistor R_{D2} is considered and designed for output voltage requirement, so feedback resistor R_{D2} is usually ignored in AC analysis. In general, the OTA type-III compensator bode plot cannot be directly simulated by SIMPLIS or HSpice without initial conditions, so the simulation circuit usually uses inductor L_{AC} and capacitor C_{AC} to generate the initial conditions. Then, the system's bode plot can be simulated.

From Fig. 2.26, a transfer function for OTA type-III compensator $G_{COMP}(s)$ can be obtained by Eqs. (2.70)–(2.73). Three poles and two zeros are used for AC analysis, as shown in Fig. 2.27. The DC gain of the OTA type-III compensator depends on gain G_M of OTA, output equivalent resistor R_O, and feedback resistor gain by Eq. (2.71). Thus, the DC gain of the OTA type-III compensator is different from the DC gain of the OTA type-II compensator. Pole S_{P0} is an initial pole, and this pole is located at a low frequency.

Fig. 2.26 Schematic of the OTA type-III compensator for AC analysis

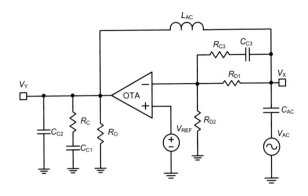

Fig. 2.27 Pole-Zero location of OTA type-III compensator for AC analysis

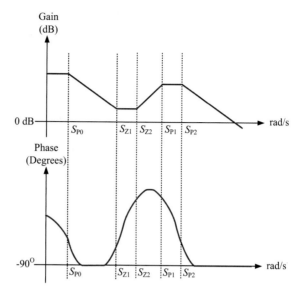

Regarding the location of pole S_{P1} and zero S_{Z1}, zero S_{Z1} should be designed smaller than pole S_{P1}. The OTA type-III compensator may increase the system loop stability, so capacitor C_{C1} should be designed much larger than capacitor C_{C2} to ensure optimal system loop stability. Moreover, pole S_{P1} is designed at a high frequency to decay the high-frequency noise, so pole S_{P1} can increase the system loop to achieve good noise immunity. Regarding the location of pole S_{P2} and zero S_{Z2}, zero S_{Z2} must be formed smaller than pole S_{P2} by Eqs. (2.75)–(2.76). Several real case designs for the OTA type-III compensator are used to simply design the location of pole S_{P2} and zero S_{Z2}. In general, assuming that feedback resistors R_{D1} and R_{D2} are much larger than resistor R_{C3}, the location of pole S_{P2} and zero S_{Z2} for transfer function $G_{COMP}(s)$ can be obtained and rewritten for the OTA type-III compensator by Eqs. (2.77)–(2.78). The location of pole S_{P2} and zero S_{Z2} depends on capacitor C_{C3} and feedback resistors R_{D1} and R_{D2}, so the location between pole S_{P2} and zero S_{Z2} depends on the output voltage. In this case, capacitor C_{C3} can be defined as a feedforward capacitor. A feedforward capacitor is widely used to parallel feedback resistor R_{D1} and to increase the phase margin for the OPA type-II compensator and OTA type-II compensator.

If the output voltage is high and the location between pole S_{P2} and zero S_{Z2} is large, then better phase degrees can be obtained. Thus, if the output voltage is low and the location between pole S_{P2} and zero S_{Z2} is small, zero S_{Z2} cannot contribute to the phase degrees. For this reason, if the output voltage is low, using the OTA type-III compensator instead of OTA type-II compensator is not good because extra cost is needed for two passive components for R_{C3} and C_{C3} to achieve the same performance in phase margin.

$$G_{COMP}(s) = \frac{\hat{V}_Y}{\hat{V}_X} = G_M \cdot R_O \cdot \frac{R_{D2}}{R_{D1} + R_{D2}} \cdot \frac{(1 + s/S_{Z1}) \cdot (1 + s/S_{Z2})}{(1 + s/S_{P0}) \cdot (1 + s/S_{P1}) \cdot (1 + s/S_{P2})}$$

(2.70)

$$G_{COMP}(0) = \frac{\hat{V}_Y}{\hat{V}_X} = G_M \cdot R_O \cdot \frac{R_{D2}}{R_{D1} + R_{D2}}$$

(2.71)

$$S_{P0} = \frac{1}{C_{C1} \cdot R_O}$$

(2.72)

$$S_{P1} = \frac{1}{C_{C2} \cdot R_C}$$

(2.73)

$$S_{Z1} = \frac{1}{C_{C1} \cdot R_C}$$

(2.74)

$$S_{P2} = \frac{1}{C_{C3} \cdot \left[\frac{(R_{D1} \cdot R_{D2} + R_{C3} \cdot R_{D2} + R_{D1} \cdot R_{C3})}{(R_{D1} + R_{D2})} \right]}$$

(2.75)

$$S_{Z2} = \frac{1}{C_{C3} \cdot (R_{C3} + R_{D1})}$$

(2.76)

$$S_{P2} = \frac{1}{C_{C3} \cdot \left(\frac{R_{D1} \cdot R_{D2}}{R_{D1} + R_{D2}} \right)}$$

(2.77)

$$S_{Z2} = \frac{1}{C_{C3} \cdot R_{D1}}$$

(2.78)

As shown in Fig. 2.27, the OTA type-III compensator has two zeros to increase the phase degrees, and the maximum phase degrees reach 180°. Thus, the OTA type-III compensator is widely used as a voltage-mode control circuit for the buck converter or voltage-mode control circuit for the boost converter. If open-loop control-to-output has two poles at the dominant frequency, designers or users may apply the OTA Type-III compensator, and the closed-loop loop gain may suffer from a bad phase margin with a large bandwidth or bad noise immunity. Pole S_{P2} cannot be removed from transfer function $G_{COMP}(s)$ for the OTA type-III compensator because pole S_{P2} does not depend on resistor R_{C3} only; it depends on feedback resistors R_{D1} and R_{D2}, so the OTA type-III compensator cannot achieve two poles and two zeros, similar to the OPA type-III compensator.

2.4   Experimental Results

MathCAD predictions, SIMPLIS simulation results, and experimental verifications were conducted to determine the feasibility and performance of the proposed the current-mode control circuit with OTA Type-II compensator for buck converter. The specifications are as follows:

(1)   Input DC voltage range (V_{IN}): 12 V
(2)   Output DC voltage range (V_{OUT}): 3.3 V
(3)   Output load current (I_{OUT}): 3A
(4)   Switching frequency (F_S): 350 kHz
(5)   Feedback resistors (R_{D1}, R_{D2}): 25.7 and 10 kΩ
(6)   Main inductor (L): 10 μH
(7)   Output capacitors (C_O): 22 μF/25 V * 2 (R_{CO}: 10 mΩ)
(8)   Reference voltage (V_{REF}): 0.925 V
(9)   Fixed ramp voltage (V_{RAMP}): 0.507 V
(10) Current Sense Ratio (R_i): 0.2 A/V
(11) OTA Type-II compensator circuit: $G_M = 1.25\,\text{mV/A}$; $R_O = 200\,\text{M}\Omega$; $R_C = 5.9\,\text{k}\Omega$; $C_{C1} = 6.2\,\text{nF}$; $C_{C2} = 158\,\text{pF}$.

Figure 2.28 shows the circuit structure of the current-mode control circuit with the OTA type-II compensator for the buck converter. Control-to-output transfer function $G_{CTO}(s)$ for the current-mode control circuit of the buck converter has three poles and one zero. $F_P(s)$ has one pole and one zero. The pole of $F_P(s)$ is located at the dominant low-frequency characteristics of the system by Eqs. (2.19)–(2.20).

MathCAD predictions of the bode plots of control-to-output, feedback resistor, and compensator are shown in Fig. 2.29. The red-colored curve represents the OTA type-II compensator bode plot, the blue-colored curve represents the control-to-output bode plot, and the brown-colored curve represents the feedback resistors' bode plot.

Fig. 2.28 Circuit structure of current-mode control circuit with OTA type-II compensator for buck converter

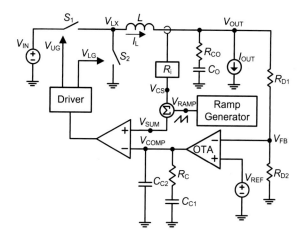

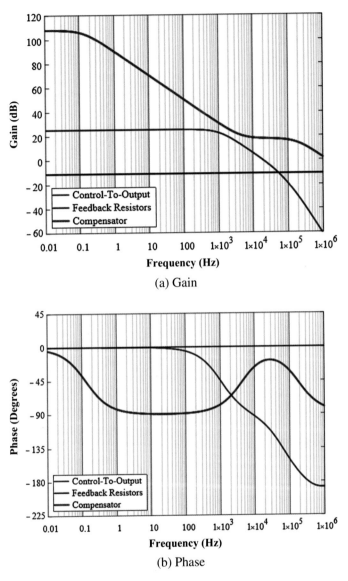

(a) Gain

(b) Phase

Fig. 2.29 MathCAD predictions of the bode plots of control-to-output, feedback resistor, and compensator

Based on the specifications, pole S_P is located at 4.3 kHz by Eq. (2.20), and zero S_{Z1} is located at 723 kHz by Eq. (2.21). Zero S_{Z1} depends on output capacitor C_O and ESR R_{CO} of the output capacitor, so zero S_{Z1} cannot contribute to the phase degrees because the output capacitor uses a low ESR R_{CO} of the ceramic capacitor.

The transfer function for OTA type-II compensator $G_{COMP}(s)$ obtained by Eqs. (2.54)–(2.58) gives two poles and one zero. The DC gain of the OTA type-II compensator depends on gain G_M of OTA and output equivalent resistor R_O. Based on the specifications, the DC gain of the OTA type-II compensator is 108 dB by Eq. (2.55). The DC gain of the OTA type-II compensator can be designed to achieve optimal bandwidth. In general, the optimal bandwidth is equal to a range between $0.1 * F_S$ and $0.15 * F_S$ when switching frequency F_S is smaller than 1 MHz. If switching frequency F_S is larger than or equal to 1 MHz and the optimal bandwidth is close to $0.1 * F_S$. A low bandwidth can obtain a slow load transient response, and a large bandwidth can result in poor noise immunity or unstable control loop. In addition, the DC gain of the OTA type-II compensator can be designed in consideration of control-to-output transfer function $G_{CTO}(s)$ and feedback resistor transfer function $G_{FB}(s)$.

With regard to OTA type-II compensator $G_{COMP}(s)$, pole S_{P0} is an initial pole and located at a low frequency. Based on the specifications, pole S_{P0} of the OTA type-II compensator is located at 0.127 Hz by Eq. (2.56). The location of pole S_{P0} depends on the output equivalent resistor R_O, and pole S_{P0} can be similar to an initial pole. Regarding the location of pole S_{P1} and zero S_{Z1}, zero S_{Z1} should be designed smaller than pole S_{P1}. The OTA type-II compensator may increase the system loop stability, so capacitor C_{C1} should be much larger than capacitor C_{C2} to ensure optimal system loop stability. To cancel pole S_P of the control-to-output transfer function $G_{CTO}(s)$, zero S_{Z1} of the OTA type-II compensator can be designed with the same location as pole S_P at 4.3 kHz. Moreover, pole S_{P1} can be designed at a high frequency to decay the high-frequency noise, so pole S_{P1} can increase the system loop to achieve good noise immunity. Pole S_{P1} is located at a half of switching frequency to ensure that pole S_{P1} does not affect the closed-loop loop gain.

Regarding the feedback resistors' transfer function $G_{FB}(s)$ by Eq. (2.79), this equation is without any pole and zero, so $G_{FB}(s)$ is a DC gain with 0 phase degree. Based on the specifications, the DC gain of the feedback resistors is −11.05 dB.

$$G_{FB}(s) = \frac{V_{REF}}{V_{OUT}} = \frac{R_{D2}}{R_{D1} + R_{D2}} \qquad (2.79)$$

Figure 2.30 shows the perturbation injection circuit at the output voltage terminal to measure the bode plot of closed-loop loop gain. If feedback resistors R_{D1} and R_{D2} are designed into the integrated circuit, closed-loop loop gain cannot be measured by a network or spectrum analyzer. However, if feedback resistors R_{D1} and R_{D2} are placed on the evaluation board to determine the output voltage, closed-loop loop gain can be measured by a network or spectrum analyzer. An AC perturbation signal V_{AC} as the input signal can be generated by a network or spectrum analyzer. V_{AC} is a sinusoidal waveform, and the frequency of the sinusoidal waveform depends on the measurement frequency. In general, the measurement frequency by a network or spectrum analyzer is between 1 kHz and 1 MHz.

Fig. 2.30 Perturbation
injection circuit at the output
voltage terminal to measure
the bode plot of closed-loop
loop gain

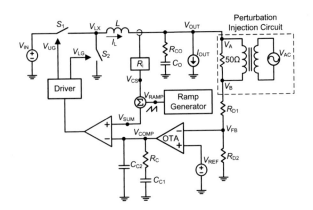

AC perturbation signal V_{AC} can be injected by a transformer, and the bandwidth of the transformer should cover a range between 1 kHz and 1 MHz to obtain a correct closed-loop loop gain bode plot. To achieve impedance matching, the output terminal should be placed at 50 Ω parallel to the transformer. The two output terminals from the network or spectrum analyzer should be connected to terminals V_A and V_B individually.

Figure 2.31 compares the MathCAD predictions, simulation results, and experimental results for the closed-loop loop gain bode plot of the current-mode control circuit with the OTA type-II compensator for the buck converter. The red-colored solid line represents the MathCAD predictions for the closed-loop loop gain bode plot, the blue-colored dotted line represents the SIMPLIS simulation results for the closed-loop loop gain bode plot, and the green-colored dashed line represents the measurement results for the closed-loop loop gain bode plot. The MathCAD predictions and SIMPLIS simulation results are similar to the experimental results.

The MathCAD predictions, SIMPLIS simulation results, and experimental results confirm that the current-mode control circuit with the OTA type-II compensator for the buck converter can significantly avoid the subharmonic issue. The bandwidth of closed-loop loop gain is equal to $0.1 * F_S$, and the bandwidth is 35 kHz. The phase margin is 50°, and the system control loop is stable. The gain margin cannot be measured in Fig. 2.31.

To further examine the advantages and superiority of the LED driver for the boost converter. LED applications for color and white high-brightness LEDs are expanding into markets, similar to smartphones to LCD TVs, architectural lighting, and general lighting. LEDs are connected in series in the output terminal, which requires a higher driving voltage level. However, most portable devices are powered by single lithium-ion battery packs, with voltage levels ranging from 2.5 to 4.2 V. Therefore, an LED driver in a portable device must step up the voltage level to drive the high-voltage LED string. Thus, a portable device needs a boost converter, not a buck converter, to drive the LED.

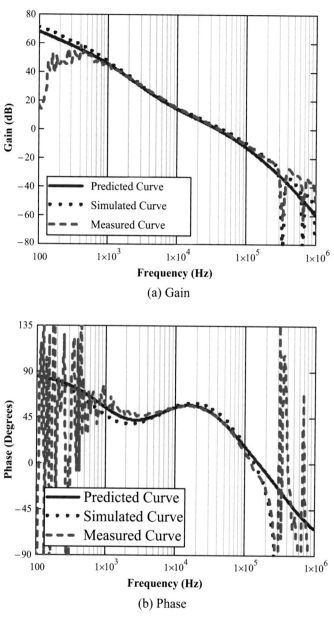

Fig. 2.31 MathCAD predictions, simulation results, and experimental results for the closed-loop loop gain bode plot of current-mode control circuit with OTA type-II compensator for buck converter

MathCAD predictions, SIMPLIS simulation results, and experimental verifications were conducted to determine the feasibility and performance of the proposed LED driver of the current-mode control circuit with OTA Type-II compensator for boost converter. The specifications are as follows:

(1) Input DC voltage range (V_{IN}): 3.7 V
(2) Output DC voltage range (V_{OUT}): 34 V (10 WLEDs)
(3) Output load current (I_{OUT}): 20 mA
(4) Switching frequency (F_S): 1 MHz
(5) Feedback resistor (R_{D2}): 5.2 Ω
(6) Main inductor (L): 22 μH
(7) Output capacitors (C_O): 1 μF/50 V * 1 (R_{CO}: 10 mΩ)
(8) Reference voltage (V_{REF}): 0.104 V
(9) Fixed ramp voltage (V_{RAMP}): 0.75 V
(10) Current Sense Ratio (R_i): 0.526 A/V
(11) OTA Type-II compensator circuit: $G_M = 1$ mV/A; $R_O = 270$ MΩ; $R_C = 99$ kΩ; $C_{C1} = 8.53$ nF; $C_{C2} = 11$ pF.

LED is similar to a diode in terms of behavior, so if the output loading is changed from pure resistors to LED, and the control loop must be designed to regulate output current I_{OUT} and not output voltage V_{OUT}. Based on the specifications, reference voltage V_{REF} is 0.104 V, so feedback voltage V_{FB} should be equal to 0.104 V, as shown in Fig. 2.32. Feedback resistor R_{D2} is designed to regulate output current I_{OUT}, which is equal to 20 mA, so feedback resistor R_{D2} is 5.2 Ω. Given that the output loading is 10 WLEDs and the control loop regulates output current I_{OUT} at 20 mA, the LEDs are under constant current, which is different from the scenario in a pure resistor. Moreover, LEDs do not need variation current, so the LED driver of the current-mode control circuit with the OTA type-II compensator for the boost converter has no load transient response requirements. This is the reason the current-mode control circuit with the OTA type-II compensator for the boost converter for LED applications is without any loop stability concern.

Figure 2.32 shows the circuit structure LED driver of the current-mode control circuit with OTA Type-II compensator for boost converter. The control-to-output transfer function $G_{CTO}(s)$ for current-mode control circuit for boost converter gives three poles and two zeros. $F_P(s)$ gives one pole, one zero, and one RHP zero by Eqs. (2.34)–(2.38). The pole of $F_P(s)$ is located at the dominant low frequency characteristics of the system by Eqs. (2.34)–(2.35). Based on one pole at the dominant low frequency, the OTA Type-II compensator or the OPA Type-II compensator is suitable to implement with current-mode control circuit for boost converter.

MathCAD predictions of the bode plots of control-to-output, feedback resistor, and compensator are shown in Fig. 2.33. The red-colored curve represents the OTA type-II compensator bode plot, the blue-colored curve represents the control-to-output bode plot, and the brown-colored curve represents the feedback resistors' bode plot.

Fig. 2.32 Circuit structure of LED driver of the current-mode control circuit with OTA type-II compensator for boost converter

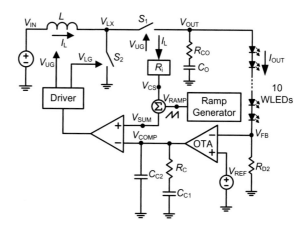

Based on the specifications, the DC gain curves of the open-loop control-to-output bode plot is equal to 41.8 dB by Eq. (2.30) using MathCAD. Pole S_P of control-to-output is located at 187.2 Hz by Eq. (2.32). Based on this operation condition using the 1 μF output capacitor, the first zero S_{Z1} is located at 13 MHz by Eq. (2.33). The second zero S_{Z2} is located at 145.6 kHz by Eq. (2.34). In this operation condition, the open-loop control-to-output bode plot shows one pole at the dominant low-frequency characteristics of the system only, so the current-mode control is more stable than the voltage-mode control with one dominant pole only.

The transfer function for the OTA type-II compensator $G_{COMP}(s)$ by Eqs. (2.54)–(2.58) gives two poles and one zero. The DC gain of the OTA type-II compensator depends on gain G_M of OTA and output equivalent resistor R_O. Based on the specifications, the DC gain of the OTA type-II compensator is 108.63 dB by Eq. (2.55). The DC gain of the OTA type-II compensator can be designed to achieve optimal bandwidth with the RHP zero limitation. Based on the RHP zero limitation, the optimal bandwidth may not reach a range between 0.1 * F_S and 0.15 * F_S when switching frequency F_S is smaller than 1 MHz, so designing the optimal bandwidth of the boost converter to be larger than that of the buck converter is difficult. In addition, the DC gain of the OTA type-II compensator can be designed considering control-to-output transfer function $G_{CTO}(s)$ and feedback resistor transfer function $G_{FB}(s)$.

In OTA type-II compensator $G_{COMP}(s)$, pole S_{P0} is an initial pole and located at a low frequency. Based on the specifications, pole S_{P0} of the OTA type-II compensator is located at 0.07 Hz by Eq. (2.56). The location of pole S_{P0} depends on output equivalent resistor R_O. Pole S_{P0} can be similar to an initial pole. Regarding the location of pole S_{P1} and zero S_{Z1}, zero S_{Z1} should be designed smaller than pole

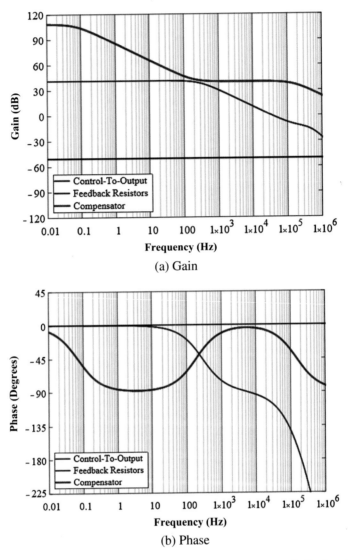

Fig. 2.33 MathCAD predictions of the bode plots of control-to-output, feedback resistor, and compensator

S_{P1}. The OTA type-II compensator may increase the system loop stability, so capacitor C_{C1} should be much larger than capacitor C_{C2} to ensure optimal system loop stability. To cancel pole S_P of the control-to-output transfer function $G_{CTO}(s)$, zero S_{Z1} of the OTA type-II compensator can be designed with the same location as

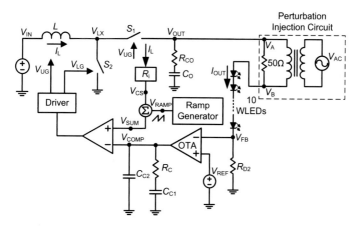

Fig. 2.34 Perturbation injection circuit at the output voltage terminal to measure the bode plot of closed-loop loop gain

pole S_P at 187.2 Hz. Moreover, pole S_{P1} is designed at a high frequency to decay high-frequency noise, so pole S_{P1} can increase the system loop to achieve good noise immunity. Pole S_{P1} is located at 145 kHz to ensure that pole S_{P1} does not affect the closed-loop loop gain.

Regarding the feedback resistors' transfer function $G_{FB}(s)$ by Eq. (2.79), this equation is without any pole and zero, so $G_{FB}(s)$ is a DC gain with 0 phase degree. Based on the specifications, the DC gain of the feedback resistors is −50.3 dB.

Figure 2.34 shows the perturbation injection circuit at the output voltage terminal to measure the bode plot of closed-loop loop gain. The output loading of LEDs should be placed on the evaluation board and feedback resistor R_{D2} to determine the output current. Closed-loop loop gain can be measured by a network or spectrum analyzer. AC perturbation signal V_{AC} as the input signal can be generated by a network or spectrum analyzer. AC perturbation signal V_{AC} is a sinusoidal waveform, and the frequency of this sinusoidal waveform depends on the measurement frequency. In general, the measurement frequency by a network or spectrum analyzer is between 1 kHz and 1 MHz.

AC perturbation signal V_{AC} can be injected by a transformer, and the bandwidth of the transformer should cover a range between 1 kHz and 1 MHz to obtain a correct closed-loop loop gain bode plot. To achieve impedance matching, the output terminal should be placed at 50 Ω parallel to the transformer. The two output terminals from the network or spectrum analyzer should be connected to terminals V_A and V_B individually.

If the perturbation injection circuit is inserted into at the output voltage terminal with output loading of LEDs, a problem would arise because the perturbation injection circuit should be placed at the high-impedance terminal. The output

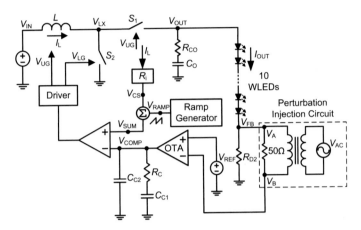

Fig. 2.35 Perturbation injection circuit at the input terminal of the OTA type-II compensator to measure the bode plot of closed-loop loop gain

loading of LEDs is similar to the current source, so the output voltage terminal is a low-impedance terminal, not a high-impedance terminal.

For this reason, the perturbation injection circuit should be placed at the input terminal of the OTA type-II compensator because this input terminal is a high-impedance terminal and a path of closed-loop loop gain. Thus, the perturbation injection circuit at the input terminal of the OTA type-II compensator can measure the correct bode plot of closed-loop loop gain, as shown in Fig. 2.35.

Figure 2.36 compares the MathCAD predictions, simulation results, and experimental results for the closed-loop loop gain bode plot of the current-mode control circuit with the OTA type-II compensator for the boost converter. The red-colored solid line represents the measurement results for the closed-loop loop gain bode plot, the blue-colored dotted line represents the MathCAD predictions for the closed-loop loop gain bode plot, and the green-colored dashed line represents the SIMPLIS simulation results for the closed-loop loop gain bode plot. The MathCAD predictions and SIMPLIS simulation results are similar to the experimental results. The MathCAD predictions, SIMPLIS simulation results, and experimental results confirm that the current-mode control circuit with the OTA type-II compensator for the boost converter can significantly prevent the subharmonic issue. In addition, the bandwidth of closed-loop loop gain hardly reaches $0.1 * F_S$, and the bandwidth is 65 kHz. The phase margin is 46°, and the system control loop is stable. The gain margin cannot be measured in Fig. 2.36.

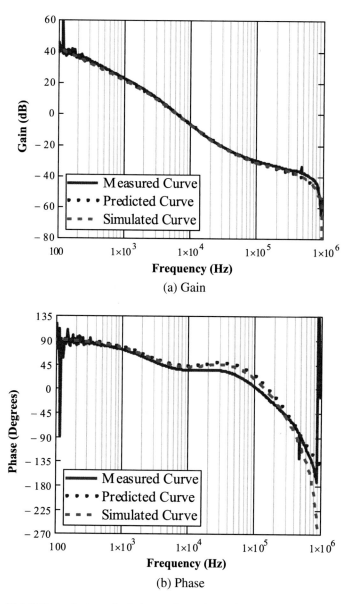

(a) Gain

(b) Phase

Fig. 2.36 MathCAD predictions, simulation results, and experimental results for the closed-loop loop gain bode plot of current-mode control circuit with OTA type-II compensator for boost converter

2.5 Summary

Experimental results, SIMPLIS simulation results, and MathCAD predictions confirm that closed-loop loop gain with the well compensation design of the current-mode control circuit with OTA Type-II compensator for buck converter does not only maintain system stability, but also significantly prevents subharmonic. Moreover, the closed-loop loop gain of the proposed LED driver of the current-mode control circuit with OTA Type-II compensator for boost converter can be verified by experimental results, SIMPLIS simulation results, and MathCAD predictions to achieve the good loop stability and to prevent subharmonic with the optimal compensation design.

References

1. C.J. Chen, D. Chen, C.W. Tseng, C.T. Tseng, Y.W. Chang, K.C. Wang, A novel ripple-based constant on-time control with virtual inductor current ripple for buck converter with ceramic output capacitors, in *Proceedings of IEEE Applied Energy Conversion Congress and Exposition Conference* (2011), pp. 1244–1250
2. W.H. Ki, Signal flow graph in loop gain analysis of DC-DC PWM CCM switching converters. IEEE Trans. Circuits Syst. Part I **45**(6), 644–654 (1998)
3. W.H. Ki, Analysis of subharmonic oscillation of fixed-frequency current-programming switch mode power converters. IEEE Trans. Circuits Syst. Part I **45**(1), 104–108 (1998)
4. A.D. Schoenfeld, Y. Yu, ASDTIC control and standardized interface circuits applied to buck, parallel and buck-boost DC-to-DC power converters, in *NASA, Washington, DC, NASA Rep. NASA* CR-121106, Feb 1973
5. C.W. Deisch, Switching control method changes power converter into a current source, in *Proceedings of IEEE Power Electronics Specialists Conference* (1978), pp. 300–306
6. P.L. Hunter, Converter circuit and method having fast responding current balance and limiting, U.S. Patent 4 002 963, 1 Nov 1977
7. L.H. Dixon, Average current-mode control of switching power supplies, in *Proceedings of Unitrode Power Supply Design Seminar Handbook* (1990), pp. 5.1–5.14
8. N. Mohan, Power electronics circuits: an overview, in *Proceedings of IEEE Industrial Electronics Society Conference* (1988), pp. 522–527
9. N. Mohan, W.P. Robbins, P. Imbertson, T.M. Undeland, R.C. Panaitescu, A.K. Jain, P. Jose, T. Begalke, Restructuring of first courses in power electronics and electric drives that integrates digital control. IEEE Trans. Power Electron. **18**, 429–437 (2003)
10. R.D. Middlebrook, S. Cuk, A general unified approach to modeling switching-converter power states, in *Proceedings of IEEE Power Electronics Specialists Conference* (1976), pp. 73–86
11. D.Y. Chen, H.A. Owen, T.G. Wilson, Computer Aided design and graphics applied to the study of inductor-energy-storage DC-to-DC electronic power converters. IEEE Trans. Aerosp. Electron. Syst. **AES-9**, 585 (1973)
12. P. Burger, Analysis of a class of pulse modulated DC-to-DC power converters. IEEE Trans. Ind. Electron. Control Instrum. **IECI-22**, 104 (1975)
13. W.H. Lau, H. Chung, C.M. Wu, N.K. Poon, Realization of digital audio amplifier using zero-voltage-switched PWM power converter. IEEE Trans. Circuits Syst. Part I **47**(3), 303–311 (2000)

14. R.B. Ridley, A new, continuous-time model for current-mode control. IEEE Trans. Power Electron. **6**, 271–280 (1991)
15. R.B. Ridley, A new continuous-time model for current-mode control with constant on-time, constant off-time, and discontinuous conduction mode, in *Proceedings of IEEE Power Electronics Specialists Conference* (1990), pp. 382–389
16. J. Li, F.C. Lee, New modeling approach and equivalent circuit representation for current-mode control. IEEE Trans. Power Electron. **25**, 1218–1230 (2010)
17. J. Li, Current-mode control: Modeling and its digital application, Ph.D. dissertation, Bradley Department of Electrical and Computer Engineering, Virginia Polytechnic Institute State University, Blacksburg, VA, USA, 2009
18. R.W. Erickson, D. Maksimovic, *Fundamentals of Power Electronics* (Kluwer, Norwell, MA, 2001)
19. W.W. Chen, J.F. Chen, T.J. Liang, L.C. Wei, W.Y. Ting, Designing a dynamic ramp with an invariant inductor in current-mode control for an on-chip Buck converter. IEEE Trans. Power Electron. **29**(2), 750–758 (2014)
20. W.W. Chen, J.F. Chen, T.J. Liang, L.C. Wei, W.Y. Ting, Dynamic Ramp with the invariant inductor in current-mode control for Buck converter, in *Proceedings of IEEE APEC* (2013), pp. 1244–1249
21. W.W. Chen, J.F. Chen, T.J. Liang, Dynamic Ramp control in current-mode adaptive on-time control for Buck converter on chip, in *Proceedings of IEEE Future Energy Electronics Conference and ECCE Asia (IFEEC 2017 - ECCE Asia)* (2017), pp. 280–285
22. S. Robert, H. Anatole, Understanding and applying current-mode control theory, in *Power Electronics Technology Exhibition and Conference* (2007)
23. V. Vorperian, Simplified analysis of PWM converters using model of PWM switch-I. Continuous conduction mode. IEEE Trans. Aerosp. Electron. Syst. **26**(3), 490–496 (1990)
24. V. Vorperian, Simplified analysis of PWM converters using model of PWM switch-II. Discontinuous conduction mode. IEEE Trans. Aerosp. Electron. Syst. **26**(3), 490–496 (1990)
25. F. Tian, S. Kasemsan, I. Batarseh, An adaptive slope compensation for the single-stage inverter with peak current-mode control. IEEE Trans. Power Electron. **26**(10), 2857–2862 (2011)
26. Z. Zansky, Current mode converter with controlled slope compensation, U.S. Patent 4 837 495, 6 June 1989
27. L. Yanming, L. Xinquan, C. Fuji, Y. Bing, J. Xinzhang, An adaptive slope compensation circuit for buck DC-DC converter, in *Proceedings of International Conference on ASIC*, Oct 2007, pp. 608–611

Chapter 3
Designing a Dynamic Ramp with Invariant Inductor in Current-Mode Control Circuit for Buck Converter

3.1 Challenges for Wide Input Voltage Range

Switching DC–DC converters are widely used in the field of power conversion due to their high efficiency and high power density. The popularity of smart phones, tablet computer, notebook computer, and similar portable battery-powered electronic devices is increasing the demand for extended battery life, thereby driving research toward high efficiency switching DC–DC converters.

In a DC–DC Buck converter, the current-mode control circuit is widely used to achieve better transient response [1–13]. The current-mode control circuit contains two feedback signals. The output voltage is fed into an error amplifier to generate a control signal. The inductor current is sampled and compared with the control signal to control the peak inductor current, thus regulating the output voltage to the specified level. The inductor current feedback causes the control to have a better transient response. However, it also produces a subharmonic issue when the duty cycle is larger than 50%.

A ramp compensation signal is always added into the control loop to prevent subharmonic [1–5, 11–16], as shown in Fig. 3.1. The ramp compensation is designed according to the inductor current ripple determined by the input voltage, output voltage, and inductance. A general control IC often imbeds a fixed ramp into the control loop. However, this kind of chip is usually designed for a wide input and output range, thus resulting in a varied inductor current ripple in the entire operation range. Meeting the fixed ramp compensation requires that the inductance is changed to maintain the same current ripple. Thus, the compensation circuit should also be modified to keep system bandwidth, gain margin, and phase margin. This modification is not convenient for the user and designer.

Figure 3.2 shows the fixed ramp sampling the output voltage in current-mode control circuit for buck converter. S_1 and S_2 are the switches, V_{UG} is upper control signal to drive the switch S_1, V_{LG} is lower control signal to drive the switch S_2, L is the output inductor, and R_{CO} is the ESR of the output capacitor C_O. The diagram

© Springer Nature Singapore Pte Ltd. 2018
W.-W. Chen and J.-F. Chen, *Control Techniques for Power Converters with Integrated Circuit*, Power Systems,
https://doi.org/10.1007/978-981-10-7004-4_3

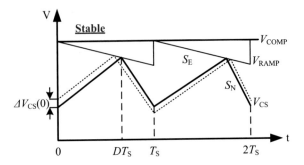

Fig. 3.1 Ramp compensation to prevent subharmonic

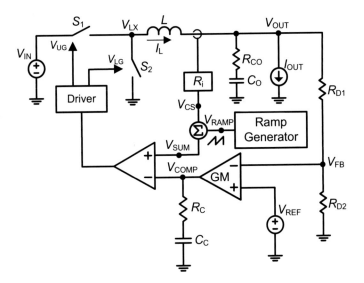

Fig. 3.2 Fixed ramp in current-mode control circuit for buck converter

shows that the output voltage ripple, which includes inductor current information, can be directly used as the ramp for modulation. The current source I_{OUT} is the output load current, and R_{D1} and R_{D2} are the feedback resistors to determine the output voltage. The feedback voltage V_{FB} and the reference voltage V_{REF} are built inside the IC. The ramp generator is always added into the control loop to prevent subharmonic. R_C and C_C are the compensator parameters to adjust the systemψloop stability.

The input voltage rang for low voltage buck is during 2.5–6 V. The input voltage range for low voltage buck is are optimized for running off single cell Li-Ion batteries, but can also be used for supplies running from 5 V rails as well. The low voltage buck can usually regulate the different output voltage rails like 1.2, 1.8, 2.5, and 3.3 V. The low voltage buck is difficult to regulate 5 V output voltage,

Table 3.1 Operation conditions in current-mode control circuit for low voltage buck converter

V_{IN} (V)	6 V	F_S (kHz)	400 kHz	L (μH)	15 μH
V_{OUT} (V)	3.3 V	R_{D1} (kΩ)	68 kΩ	C_O (μF)	22 μF
I_{OUT} (A)	2 A	R_{D2} (kΩ)	22 kΩ	R_{CO} (mΩ)	3 mΩ
V_{REF} (V)	0.8 V	R_i (A/V)	0.4		

because it is needed to depend on the $R_{DS(ON)}$ of the switch S_1 and the maximum output current rating.

Table 3.1 lists the operation conditions of current-mode control circuit for low voltage buck converter, where m_C is expressed by Eqs. (3.1) and (3.2). S_E is a slope of the ramp compensation and S_N is the inductor on-time slope and $D' = 1 - D$. Based on Table 3.1, the open-loop control-to-output bode plot with different m_C values is plotted in Fig. 3.3.

$$m_C = 1 + \frac{S_E}{S_N} \tag{3.1}$$

$$S_N = \frac{V_{IN} - V_{OUT}}{L} \cdot R_i \tag{3.2}$$

In Fig. 3.3 [1–5], the gain curves of m_C values 1 and 1.2 present no ramp compensation and small ramp compensation, respectively. Both of these values have two poles which can cause a high Q at half of the switching frequency. The two poles cause a sharp phase drop of 180° at a half of the switching frequency, thus resulting in subharmonic and an unstable system. However, if the ramp compensation is large, as presented by the gain curve of m_C is equal to 8, then the control loop has formed as a voltage-mode control, given that the ramp compensation is much larger than the inductor current. Therefore, a proper m_C value between 2 and 4 should be designed to prevent subharmonic and to ensure the stability of the system.

Most low voltage buck converters provide a recommended component selection table to maintain parameter m_C in a proper range, as listed in Table 3.2 [17, 18]. In order to meet the fixed ramp compensation requires that the inductance is changed to maintain the same current ripple when the user determines to apply different the output voltage. Thus, the compensation circuit should also be modified to keep the system stability. Based on the recommended component selection table, the inductance is a minor change at different output voltage, because the voltage drop between the input voltage and output voltage cannot result in a large value for S_N.

Table 3.3 lists the same inductance and compensator under different output voltages in current-mode control circuit for low voltage buck converter. Based on the conditions listed in Table 3.3, bode plot is constructed to illustrate system stability, as shown in Fig. 3.4. The operation of the input voltage is 3.3 V for the worst case scenario (high duty cycle). Inductance and capacitance are also selected at 1.5 μH and 22 μF, respectively, under different output voltages. Figure 3.4 shows

Fig. 3.3 Open-loop
control-to-output bode plot
with different m_C

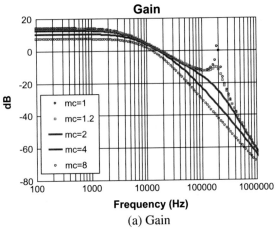

(a) Gain

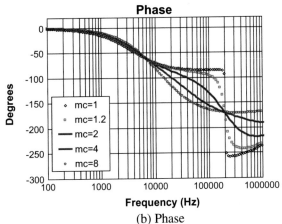

(b) Phase

Table 3.2 Recommended
component selection for low
voltage buck converter

Conditions	V_{IN} = 2.5–6 V, F_S = 1000 kHz, I_{OUT} = 2A			
IC internal	V_{RAMP} = 566 mV, V_{REF} = 827 mV			
V_{OUT} (V)	R_{D1} (kΩ)	R_{D2} (kΩ)	L (µH)	C_O (µF)
3.3	4.3	10	2.2	22
2.5	11.8	10	2.2	22
1.8	20.4	10	1.5	22
1.2	30	10	1.5	22

that the gain curve of the output voltage of 3.3 V still contains current-mode
behavior to increase the loop stability. Based on the control-to-output bode plot
results, the control-to-output cannot contain two poles under different output voltages, so the low voltage buck converter is not a big challenge for avoiding a subharmonic issue, because S_N is not a wide range variation.

Table 3.3 Same inductance and compensator under different output voltages for low voltage buck converter

Conditions	V_{IN} = 2.5–6 V, F_S = 1000 kHz, I_{OUT} = 2A			
IC internal	V_{RAMP} = 566 mV, V_{REF} = 827 mV			
V_{OUT} (V)	R_{D1} (kΩ)	R_{D2} (kΩ)	L (μH)	C_O (μF)
3.3	4.3	10	1.5	22
2.5	11.8	10	1.5	22
1.8	20.4	10	1.5	22
1.2	30	10	1.5	22

Fig. 3.4 Open-loop control-to-output bode plot with different output voltages at the input voltage 6 V for low voltage buck converter

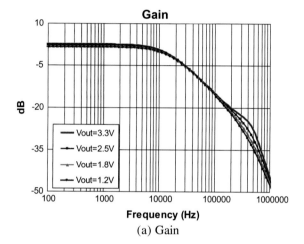

(a) Gain

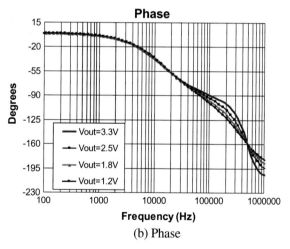

(b) Phase

The wide input voltage range for high voltage buck is during 4.75–22 V. The wide input voltage range for high voltage buck is normally used for adapter applications that run from 12 V supply rails or notebook computer applications that run from 19 V supply rails, but the wide input voltage range makes it possible to

Table 3.4 Recommended component selection for high voltage buck converter

Conditions	V_{IN} = 4.75–22 V, F_S = 400 kHz, I_{OUT} = 2A				
IC internal	V_{RAMP} = 306 mV, V_{REF} = 1222 mV				
V_{OUT} (V)	R_{D1} (kΩ)	R_{D2} (kΩ)	L (µH)	C_O (µF)
12	88.7	10	33	22
5	30	10	22	22
3.3	17	10	15	22
1.8	4.75	10	10	22

Table 3.5 Same inductance and capacitance under different output voltages for high voltage buck converter

Conditions	V_{IN} = 4.75–22 V, F_S = 400 kHz, I_{OUT} = 2A				
IC internal	V_{RAMP} = 306 mV, V_{REF} = 1222 mV				
V_{OUT} (V)	R_{D1} (kΩ)	R_{D2} (kΩ)	L (µH)	C_O (µF)
12	88.7	10	15	22
5	30	10	15	22
3.3	17	10	15	22
1.8	4.75	10	15	22

run them from 5 V rails as well. The high voltage buck can usually regulate the different output voltage rails like 12, 5, 3.3, and 1.8 V.

Most high voltage buck converters provide a recommended component selection table to maintain parameter m_C in a proper range, as listed in Table 3.4 [17, 18]. In order to meet the fixed ramp compensation requires that the inductance is changed to maintain the same current ripple when the user determines to apply different the output voltage. Thus, the compensation circuit should also be modified to keep the system stability. This modification is not convenient for the user and designer. Therefore, this proposed circuit of a novel control scheme can solve this issue.

Table 3.5 lists the same inductance and capacitance under different output voltages in current-mode control circuit for high voltage buck converter. Based on the conditions listed in Table 3.5, bode plot is constructed to illustrate system stability, as shown in Fig. 3.5. The operation of the input voltage is 14 V for the worst case scenario (high duty cycle). Inductance and capacitance are also selected at 15 µH and 22 µF, respectively, under different output voltages. Figure 3.5 shows that the gain curve of the output voltage of 12 V contains two poles which can result in a high Q at half of the switching frequency. The two poles cause a sharp phase drop of 180° at a half of the switching frequency, which means that the system suffers from a subharmonic issue at the output voltage of 12 V, so the high voltage buck converter is a big challenge for avoiding a subharmonic issue, because S_N has a wide range variation. On the other hand, the compensation circuit is difficult to design cancelling a subharmonic issue.

Fig. 3.5 Open-loop
control-to-output bode plot
with different output voltages
at the input voltage 14 V for
high voltage buck converter

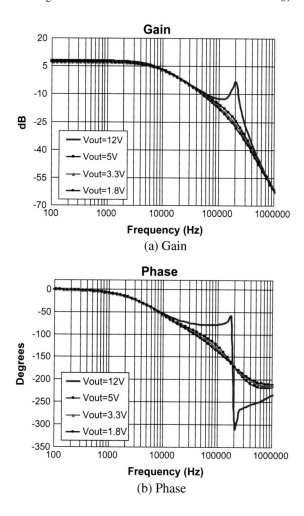

(a) Gain

(b) Phase

3.2 Dynamic Slope Compensation Design

Figure 3.6 shows the dynamic ramp sampling the output voltage in current-mode control circuit for buck converter. The ramp generator can only sample the output voltage or the output voltage and input voltage, to implement a dynamic ramp. The output voltage is changed large, and the dynamic ramp is increased to prevent the system suffering from a subharmonic issue.

Figure 3.7 shows that the dynamic ramp is dependent on the output voltage. The duty cycle D_1 is small than 50% as shown in Fig. 3.7a and the duty cycle D_2 larger than 50% as shown in Fig. 3.7b. Based on the duty cycle D_1 is smaller than the duty cycle D_2 and the slope of S_{N1} is steeper than that of S_{N2}. The ramp generator yields the new ramp compensation V_{RAMP2} instead of the original ramp compensation

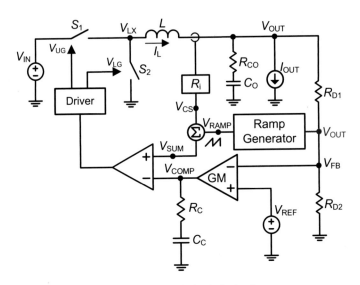

Fig. 3.6 Dynamic ramp in current-mode control circuit for buck converter

Fig. 3.7 Dynamic ramp depended on the output voltage

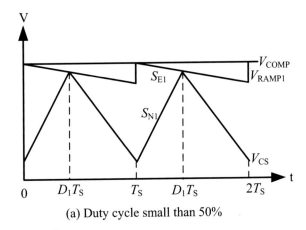

(a) Duty cycle small than 50%

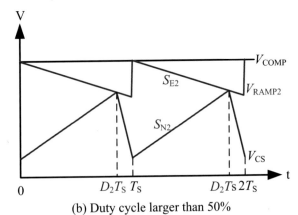

(b) Duty cycle larger than 50%

Fig. 3.8 Implementation of a dynamic ramp generator sampling the output voltage

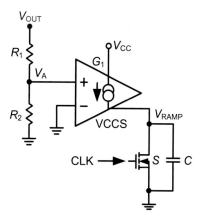

V_{RAMP1}, which has the slope of S_{E2} steeper than that of S_{E1}. Thus, ramp compensation is not fixed but dynamic and the dynamic ramp is convenient for the user and designer, because it does not change the inductor to meet different output voltages operation.

Figure 3.8 shows the implementation of the dynamic ramp generator sampling the output voltage. The voltage V_A depends on the output voltage. The voltage V_A is also an input signal for the VCCS. The output signal of VCCS charges capacitor C to generate the dynamic ramp V_{RAMP}, which can be calculated by Eq. (3.3). G_1 is the gain of the voltage-controlled current source. The switch S is controlled to connect the voltage V_{RAMP} to the ground and the control signal of CLK is usually used 1–5% of duty cycle to control the switch S.

$$V_{\mathrm{RAMP}} = \frac{G_1 \cdot \left(V_{\mathrm{OUT}} \times \frac{R_2}{R_1 + R_2}\right)}{C \cdot T_S} \tag{3.3}$$

Figure 3.9 shows the block diagram of the dynamic ramp sampling the output voltage in current-mode control circuit for buck converter. In the block diagram, the switches of the power stage are replaced with a three-terminal switching model [1–4, 10–13, 19–28], in which the ideal transformer plays the role of the average duty cycle, and the two dependent sources model the perturbation of the duty cycle. According to the block diagram, transfer functions can be derived to analyze system stability. The transfer functions of control-to-output are shown in Eqs. (3.4)–(3.9). m_C should be designed at an optimum proportion between S_N and S_E to obtain the suitable signal for S_E.

$$\frac{\hat{V}_{\mathrm{OUT}}}{\hat{V}_{\mathrm{COMP}}} \cong \frac{R_{\mathrm{OUT}}}{R_i} \cdot \frac{1}{1 + \frac{R_{\mathrm{OUT}}}{L \cdot F_S}\left[\left(1 + \frac{S_E}{S_N}\right) \cdot D' - 0.5\right]} \cdot \mathrm{Fp}(s) \cdot \mathrm{Fh}(s) \tag{3.4}$$

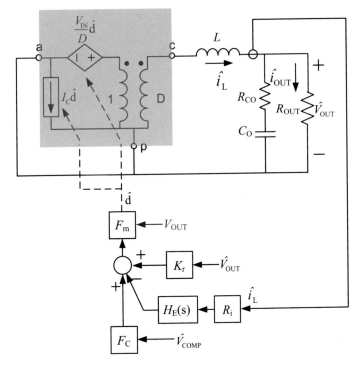

Fig. 3.9 Block diagram of the dynamic ramp sampling the output voltage in current-mode control circuit for buck converter

$$Fp(s) = \frac{1 + s \cdot C_O \cdot R_{CO}}{1 + \frac{s}{\omega_P}} \tag{3.5}$$

$$Fh(s) = \frac{1}{1 + \frac{s}{\omega_n \cdot Q_P} + \frac{s^2}{\omega_n^2}} \tag{3.6}$$

$$\omega_P \frac{1}{C_O \cdot R_{OUT}} + \frac{1}{C_O \cdot L \cdot F_S} \cdot \left[\left(1 + \frac{S_E}{S_N} \right) \cdot D' - 0.5 \right] \tag{3.7}$$

$$Q_P = \frac{1}{\pi \cdot \left[\left(1 + \frac{S_E}{S_N} \right) \cdot D' - 0.5 \right]} \tag{3.8}$$

$$\omega_n = \pi \cdot F_S \tag{3.9}$$

Based on the conditions listed in Table 3.5 for optimized m_C values, inductance and capacitance are selected at 15 µH and 22 µF, respectively, at the input voltage of 14 V and the output voltage of 12 V, under different m_C values of 3, 6, and 9, as shown in Fig. 3.10.

Fig. 3.10 Open-loop
control-to-output bode plot
with different m_C at the input
voltage 14 V and the output
voltage 12 V

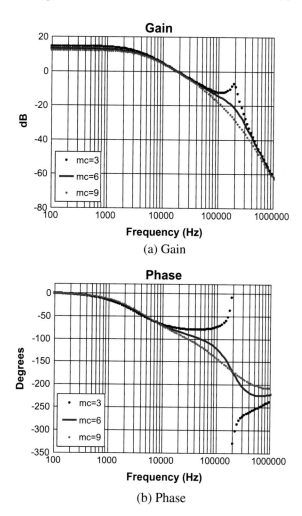

(a) Gain

(b) Phase

The gain curve of $m_C = 3$ contains a high Q at half of the switching frequency caused by two poles. The gain curve of $m_C = 9$ has a voltage mode behavior similar to having no current loop function. Finally, the gain curve of $m_C = 6$ is optimum for designing the proportion between S_N and S_E. This system is stable to operate at a high duty cycle and also uses the invariant inductor to prevent subharmonic. The dynamic ramp is designed based on the gain curve of $m_C = 6$. Ramp generator parameters G_1, R_1, R_2, and C are calculated by Eq. (3.10). Assuming that R_1 is 100 kΩ, R_2 is 10 kΩ, and C is 10 pF, then gain G_1 of VCCS is 2.6667μ.

$$\frac{G_1 \cdot (V_{OUT} \cdot \frac{R_2}{R_1 + R_2})}{C} \geq 5 \cdot \frac{(V_{IN} - V_{OUT})}{L} \cdot R_i \qquad (3.10)$$

Fig. 3.11 Open-loop
control-to-output bode plot of
the dynamic ramp sampling
the output voltage with
different output voltages at the
input voltage 14 V

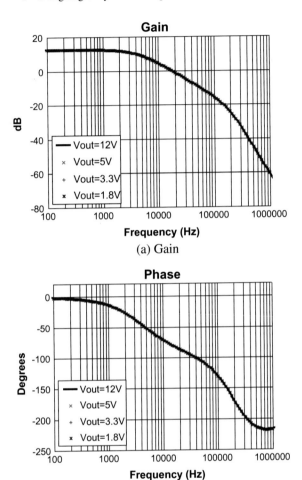

(a) Gain

(b) Phase

Figure 3.11 shows the open-loop control-to-output bode plot of the dynamic ramp sampling the output voltage with different output voltages at the input voltage 14 V. The open-loop control-to-output for buck converter using the dynamic ramp exhibits the same bode plot for the gain and phase under different output voltages. This system does not only use the invariant inductor, but also designs the same compensation circuit to meet different applications when the output voltage is changed. Therefore, the proposed dynamic ramp sampling the output voltage is very simple and easy to implement.

The ramp generator can also sample the output voltage and the input voltage to implement the dynamic ramp. The voltage droop between the input voltage and the output voltage is changed from large to small, and the dynamic ramp is increased to prevent the system suffering from a subharmonic issue. Figure 3.12 shows the

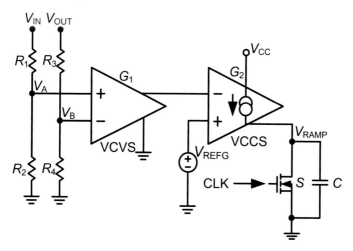

Fig. 3.12 Implementation of a dynamic ramp generator sampling the output voltage and input voltage

implementation of the dynamic ramp generator with sampling the output voltage and input voltage.

The voltage V_A depends on the input voltage. The voltage V_A is also an input signal for a VCVS. The voltage V_B depends on the output voltage. This voltage is the other input signal for a VCVS. G_1 is the gain of VCVS. G_1 is also a unity gain. The output signal of VCVS is an input signal for VCCS. G_2 is the gain of VCCS. The output signal of VCCS charges capacitor C to generate the dynamic ramp V_{RAMP}, which can be calculated by Eq. (3.11). Otherwise, the voltage signal V_{REFG} should be designed to be larger than the voltage droop between the V_A and V_B.

$$V_{RAMP} = \frac{G_2 \cdot \left[V_{REFG} - \left(V_{IN} \cdot \frac{R_2}{R_1 + R_2} - V_{OUT} \cdot \frac{R_4}{R_3 + R_4} \right) \right]}{C \cdot T_S} \quad (3.11)$$

Figure 3.13 shows the block diagram of the dynamic ramp with sampling the output voltage and the input voltage in current-mode control circuit for buck converter. Finally, the gain curve of $m_C = 6$ is also optimum for designing the proportion between S_N and S_E as shown in Fig. 3.10. This system is stable to operate at a high duty cycle and also uses the invariant inductor to prevent subharmonic. Ramp generator parameters G_2, R_1, R_2, R_3, R_4, V_{REFG}, and C are calculated by Eq. (3.11).

Figure 3.14 shows the open-loop control-to-output bode plot of the dynamic ramp sampling the output voltage and the input voltage with different output voltages at the input voltage of 14 V. According to the same bode plot, this system does not only use the invariant inductor, but also designs the same compensation circuit to meet different applications when the output voltage is changed. Therefore, the proposed dynamic ramp sampling the output voltage and the input voltage is very simple and easy to implement.

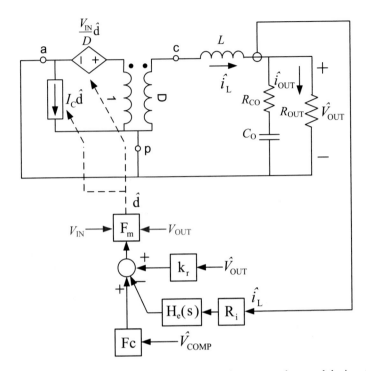

Fig. 3.13 Block diagram of the dynamic ramp sampling the output voltage and the input voltage in current-mode control circuit for buck converter

3.3 Experimental Results

MathCAD predictions, SIMPLIS simulation results, and experimental verifications were conducted to determine the feasibility and performance of the proposed dynamic ramp with the invariant inductor in current-mode control circuit for buck converter. The specifications are as follows:

1. Input DC voltage range (V_{IN}): 14 V
2. Output DC voltage range (V_{OUT}): 12 V
3. Output load current (I_{OUT}): 2 A
4. Switching frequency (F_S): 400 kHz
5. Feedback resistors (R_{D1}, R_{D2}): 115 and 13 kΩ
6. Main inductor (L): 15 μH
7. Output capacitors (C_O): 22 μF/ 25 V * 1 (R_{CO}: 3 mΩ)
8. Reference voltage (V_{REF}): 1.222 V

Fig. 3.14 Open-loop
control-to-output bode plot of
the dynamic ramp sampling
the output voltage and the
input voltage with different
output voltages at the input
voltage 14 V

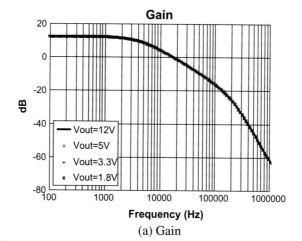

(a) Gain

(b) Phase

9. Dynamic ramp circuit (Fig. 3.8): $G_1 = 2.6667 \ \mu\text{V/A}$; $C = 10\text{p}$; $R_1 = 100 \ \text{k}\Omega$; $R_2 = 10 \ \text{k}\Omega$
10. Fixed ramp circuit: $V_{\text{RAMP}} = 0.306 \ \text{V}$
11. Compensator circuit: $R_C = 10 \ \text{k}\Omega$; $C_C = 1.5 \ \text{nF}$

Figure 3.15 compares MathCAD predictions and SIMPLIS simulation results
for the open-loop control-to-output bode plot using the dynamic ramp and the fixed
ramp. MathCAD is used for predicting results in calculating equations for the
dynamic ramp and the fixed ramp. SIMPLIS is used to simulate results for the
dynamic ramp and the fixed ramp. MathCAD prediction results are very similar to
SIMPLIS simulation results. The fixed ramp has a high Q at half of the switching
frequency caused by two poles. The two poles cause a sharp phase drop of 180° at
200 kHz, which is also located at half of the switching frequency, thus resulting in a
subharmonic issue and an unstable system.

Fig. 3.15 Comparison of
MathCAD predictions and
SIMPLIS simulation results
for open-loop
control-to-output bode plot
using the dynamic ramp and
the fixed ramp

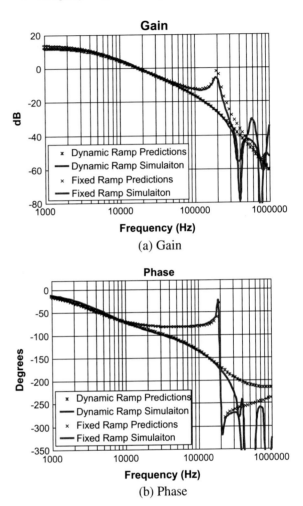

(a) Gain

(b) Phase

Figure 3.16 compares MathCAD predictions and SIMPLIS simulation results
for the closed-loop loop gain bode plot using the dynamic ramp and the fixed
ramp. In Fig. 3.15, MathCAD prediction results are very similar to SIMPLIS
simulation results. Both MathCAD prediction and SIMPLIS simulation result
verifications confirm that the proposed dynamic ramp with the invariant inductor in
current-mode control circuit for buck converter can significantly prevent subharmonic. The bandwidth of the fixed ramp for the SIMPLIS simulation results and
MathCAD predictions are both 20 kHz; whereas the bandwidth of the dynamic
ramp for SIMPLIS simulation results and MathCAD predictions are very close to
each other. The phase margin of the dynamic ramp is very close to the fixed ramp.

Figure 3.17 shows the chip layout of the control IC. The control IC was used by
TSMC 0.6 μm process and $R_{DS(ON)}$ of the main switch is 180 mΩ. The Power

Fig. 3.16 Comparison of
MathCAD predictions and
SIMPLIS simulation results
for closed-loop loop gain
bode plot using the dynamic
ramp and the fixed ramp

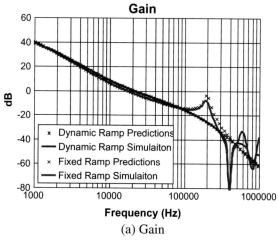

(a) Gain

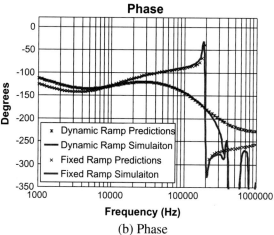

(b) Phase

MosFET, driver circuit, control circuit, and current sensor occupy the marked area
with a die size of 1250 μm × 960 μm.

Figure 3.18 shows the output voltage and the V_{UG} signal using the fixed ramp at
the input voltage of 14 V, the output voltage of 12 V, and the output load current of
2A. Experimental results are measured at the output voltage and the V_{UG} signal for
the fixed ramp. The experimental result of the output voltage ripple for the fixed
ramp is equal to 27.2 mV. The V_{UG} signal for the fixed ramp exhibits subharmonic,
thus the width of the V_{UG} signal is not the same.

Figure 3.19 shows the output voltage and the V_{UG} signal using the dynamic
ramp at the input voltage of 14 V, the output voltage of 12 V, and the output load
current of 2 A. Experimental results are measured at the output voltage and the V_{UG}
signal for the dynamic ramp. The experimental result of the output voltage ripple

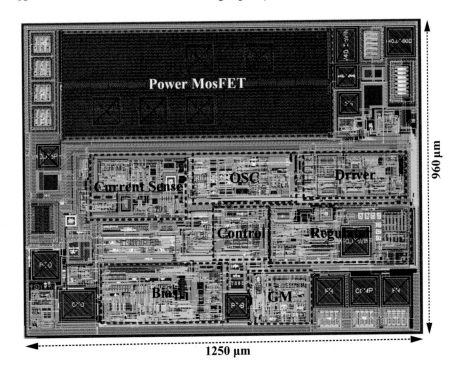

Fig. 3.17 Chip layout of the control IC

Fig. 3.18 V_{OUT} and V_{UG}
signals using the fixed ramp at
the input voltage 14 V and the
output voltage 12 V

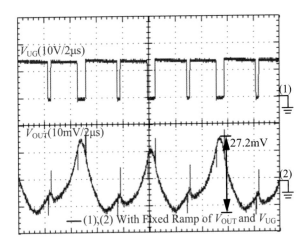

for the dynamic ramp is equal to 6.8 mV. The V_{UG} signal does not exhibit sub-harmonic and the width of the V_{UG} signal is the same.

Figure 3.20 compares of the experimental results for efficiency using the dynamic ramp and the fixed ramp at the input voltage 14 V and the output voltage

Fig. 3.19 V_{OUT} and V_{UG} signals using the dynamic ramp at the input voltage 14 V and the output voltage 12 V

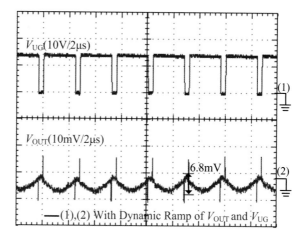

Fig. 3.20 Comparison of the experimental results for efficiency using the dynamic ramp and the fixed ramp at the input voltage 14 V and the output voltage 12 V

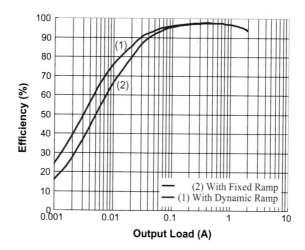

12 V in current-mode control circuit for buck converter. The current-mode control circuit using the dynamic ramp for buck converter has higher efficiency than the fixed ramp during the output load current 1 mA to 2 A, because the fixed ramp has the subharmonic resulting in the large switching loss. When the output load current has increased, the efficiency of that with the fixed ramp is very close to that with the dynamic ramp, because the conduction loss is a major factor that impacts the efficiency in this condition.

Figure 3.21 compares MathCAD predictions and experimental results for the closed-loop loop gain bode plot using the dynamic ramp and the fixed ramp. MathCAD prediction results are very similar to experimental results. Both MathCAD prediction and experimental result verification confirm that the proposed

Fig. 3.21 Comparison of
MathCAD predictions and
experimental results for
closed-loop loop gain bode
plot using the dynamic ramp
and the fixed ramp

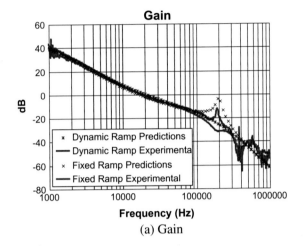

(a) Gain

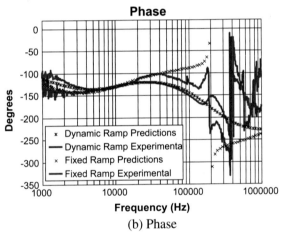

(b) Phase

dynamic ramp with the invariant inductor in current-mode control circuit for buck
converter can significantly prevent subharmonic. The gain curve of the fixed ramp
contains a high Q at 200 kHz, which is also located at half of the switching
frequency caused by two poles.

The dynamic ramp and fixed ramp both uses the same compensation parameters
to obtain MathCAD predictions and experimental results. The bandwidth of the
fixed ramp for the experimental results and MathCAD predictions are both 20 kHz;
whereas the bandwidth of the dynamic ramp for the experimental results and
MathCAD predictions are very close to each other. The phase margin of the
dynamic ramp is very close to the fixed ramp for both the experimental results and
MathCAD predictions.

3.4 Summary

Experimental results, SIMPLIS simulation results, and MathCAD predictions confirm that the proposed dynamic ramp with the invariant inductor in current-mode control circuit for buck converter does not only maintain system stability under different input/output voltages without changing the inductor and compensation circuit, but also significantly prevents subharmonic. Moreover, the proposed dynamic ramp has a very simple structure and easy-to-implement design. This system is available and useful as a current-mode control for buck converter.

References

1. R.B. Ridley, A new, continuous-time model for current-mode control. IEEE Trans. Power Electron. **6**, 271–280 (1991)
2. R.B. Ridley, A new continuous-time model for current-mode control with constant on-time, constant off-time, and discontinuous conduction mode, in *Proceedings on IEEE Power Electronics Specialists Conference* (1990), pp. 382–389
3. J. Li, F.C. Lee, New modeling approach and equivalent circuit representation for current-mode control. IEEE Trans. Power Electron. 1218–1230 (2010)
4. R.W. Erickson, D. Maksimovic, *Fundamentals of Power Electronics* (Kluwer, Norwell, MA, 2001)
5. R.D. Middlebrook, S. Cuk, A general unified approach to modeling switching-converter power states, in *Proceedings on IEEE Power Electronics Specialists Conference* (1976), pp. 73–86
6. N. Mohan, W.P. Robbins, P. Imbertson, T.M. Undeland, R.C. Panaitescu, A.K. Jain, P. Jose, T. Begalke, Restructuring of first courses in power electronics and electric drives that integrates digital control. IEEE Trans. Power Electron. **18**, 429–437 (2003)
7. N. Mohan, Power electronics circuits: An overview, in *Proceedings on IEEE Industrial Electronics Society Conference* (1988), pp. 522–527
8. W.H. Ki, Signal flow graph in loop gain analysis of dc-dc PWM CCM switching converters. IEEE Trans. Circ. Syst.-I **6**, 644–654 (1998)
9. W.H. Ki, Analysis of subharmonic oscillation of fixed-frequency current-programming switch mode power converters. IEEE Trans. Circ. Syst.-I **45**(1), 104–108 (1998)
10. J. Li, Current-mode control: modeling and its digital application, PHD thesis, Virginia Polytechnic Institute and State University, 2009
11. W.W. Chen, J.F. Chen, T.J. Liang, L.C. Wei, W.Y. Ting, Designing a dynamic ramp with an invariant inductor in current-mode control for an on-chip Buck converter. IEEE Trans. Power Electron. **29**(2), 750–758 (2014)
12. W.W. Chen, J.F. Chen, T. J. Liang, L. C. Wei, W.Y. Ting, Dynamic Ramp with the invariant inductor in current-mode control for Buck converter, in *Proceedings on IEEE Applied Power Electronics Conference and Exposition (APEC)* (2013), pp. 1244–1249
13. W.W. Chen, J.F. Chen, T.J. Liang, Dynamic Ramp control in current-mode adaptive on-time control for Buck converter on chip, in *Proceedings on IEEE Future Energy Electronics Conference and ECCE Asia (IFEEC 2017—ECCE Asia)* (2017), pp. 280–285
14. F. Tian, S. Kasemsan, I. Batarseh, An adaptive slope compensation for the single-stage inverter with peak current-mode control. IEEE Trans. Power Electron. 2857–2862 (2011)
15. Z. Zansky, Current Mode Converter with Controlled Slope Compensation. United States Patent, Patent Number: 4,837,495, Date of Patent: June 6, 1989

16. L. Yanming, L. Xinquan, C. Fuji, Y. Bing, J. Xinzhang, An adaptive slope compensation circuit for buck DC-DC converter, in *Proceeding of the International Conference on ASIC 2007*, ASICON'07 (Oct 2007) pp. 608–611
17. Richtek Tech. Corp., 2A, 22 V, 400 kHz Step-Down Converter. RT8267 Datasheet (2011)
18. Richtek Tech. Corp., 2A, 23 V, 340 kHz Synchronous Step-Down Converter. RT8294 Datasheet (2011)
19. R. Redl, N.O. Sokal, Current-mode control, five different types, used with the three basic classes of power converters: Small-signal ac and large-signal dc characterization, stability requirements, and implementation of practical circuits, in *Proceedings on IEEE Power Electronics Specialists Conference* (1985), pp. 771–785
20. N. Lakshminarasamma, M. Masihuzzaman, V. Ramanarayanan, Steady-state stability of current-mode active-clamp ZVS DC-DC converters. IEEE Trans. Power Electron. 26(5), 1295–1304 (2011)
21. B. Bryant, M.K. Kazimierczuk, Modeling the closed-current loop of PWM dc-dc converters operating in CCM with peak current-mode control. IEEE Trans. Circ. Syst.-I 52(11), 2404–2412 (2005)
22. N. Kondrath, M.K. Kazimierczuk, Control current and relative stability of peak current-mode controlled pulse-width modulated dc-dc converters without slope compensation. IET Power Electron. 3(6), 936–946 (2010)
23. M.K. Kazimierczuk, Pulse-Width Modulated DC-DC Power Converters (Wiley, 2008)
24. V. Vorperian, Simplified analysis of PWM converters using model of PWM switch-I. Continuous conduction mode. IEEE Trans. Aerosp. Electron. Syst. 26(3), 490–496 (1990)
25. V. Vorperian, Simplified analysis of PWM converters using model of PWM switch-II. Discontinuous conduction mode. IEEE Trans. Aerosp. Electron. Syst. 26(3), 490–496 (1990)
26. B. Bryant, M.K. Kazimierczuk, Open-loop power-stage transfer functions relevant to current-mode control of boost PWM converter operating in CCM. IEEE Trans. Circ. Syst.-I 52(11), 2158–2164 (2005)
27. M.M. Jovanovic, L. Huber, Small-signal modeling of nonideal magamp PWM switch. IEEE Trans. Power Electron. 14(5), 882–889 (1999)
28. Y. Yingyi, F.C. Lee, P. Mattavelli, Unified three-terminal switch model for current mode controls. IEEE Trans. Power Electron. 27(9), 4060–4070 (2012)

Chapter 4
Review of the Adaptive On-time Control Circuits for Buck Converters

4.1 Adaptive On-time Control Circuits for Buck Converters

Many novel control circuits, such as CPU and electronic devices, have been reported for power supplies to meet stringent requirements in recent years. These devices can reduce standby power loss and increase the load transient response to achieve high performance and low loss of system design. Owing to the rapid development of microprocessors, over a billion transistors have been integrated into one processor. Core static current has been increased from 20 to 100 A, and core voltage has been reduced from 2 to 0.7 V [1–3]. Moreover, CPU load transient may occur within 1 μs. These requirements pose a stringent challenge to voltage regulation (VR) [4–8], especially in the load transient response. The use of many output capacitors to reduce the voltage spike during transient is an approach that has been adopted. For future microprocessors, increasing the number of capacitors to meet even higher transient requirements will not be acceptable because of size and cost issues.

The power converter must be able to regulate its output voltage to be near constant as the load current demand varies anywhere from zero to full load, even when the change occurs in a relatively short time. A good performance of load transient response can save on output capacitor size and cost. Meanwhile, settling time and stability can be displayed in the load transient response, so power converter performance must be tested. Based on these requirements, the conventional COT control circuit is widely used in CPU applications and other electronic devices with high slew rates because of the advantages of faster load transient response and better light-load efficiency compared with the current-mode control. The first reason is that the conventional COT control circuit does not have a compensator, which delays the system loop response. The second reason is that the conventional COT control circuit is a type of PFM, so it can generate a minimum off-time at load droop, which is very useful in preventing the output voltage from dropping

© Springer Nature Singapore Pte Ltd. 2018
W.-W. Chen and J.-F. Chen, *Control Techniques for Power Converters with Integrated Circuit*, Power Systems,
https://doi.org/10.1007/978-981-10-7004-4_4

significantly. Thus, the conventional COT control circuit is suitable for use in fast load transient power supply applications.

However, the load transient response of conventional COT control circuit requires further improvement to save output capacitance, so it is used in CPU application, but CPU load transient may occur within 1 µs, so the conventional COT control circuit is also needed to add a non-linear open loop quick response circuit to improve transient response [9–21]. Further, most conventional COT controls provide voltage to CPU, try to solve this issue by setting up drop-voltage thresholds to trigger another open loop regulation mechanism, such as triggering another on-time generator or increasing the duty of its on-time generator. However, this kind of design has two major drawbacks. First, the threshold is discrete, which means it may improve transient response over a specific threshold. Second, the threshold is fixed, which cannot meet the variety of loading conditions. Moreover, if the threshold can be set by external components, it suffers another drawback of adding extra pins, which increases cost and reduces the flexibility of board design.

Our desire is to provide a method, which can maintain a switching regulator's loop stability and characteristic, and also provide a nearly real-time boost response to help the regulator trace back the correct voltage. This method can allow the output voltage drop to decide dynamically how fast or what quantity to boost the regulator without the extra setting of pins and provide board designers the flexibility to change boost speed.

Circuit diagram of the conventional COT control circuit for buck converter is as shown Fig. 4.1. S_A and S_B are the switches, L is the output inductor, and R_{CO} is the ESR of the output capacitor C_O. The current source I_{OUT} is the output load current,

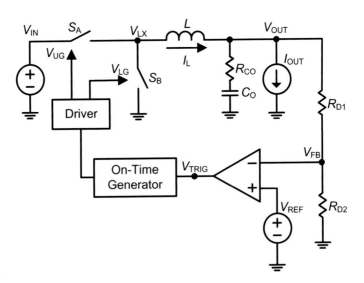

Fig. 4.1 Circuit diagram of the conventional COT control circuit for buck converter

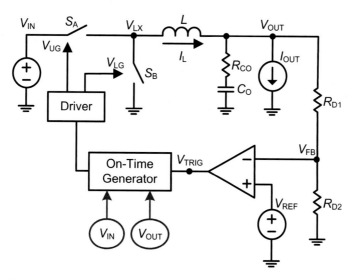

Fig. 4.2 Adaptive on-time control circuits for buck converter

and R_{D1} and R_{D2} are the feedback resistors to determine the output voltage. V_{FB} is a feedback voltage. The reference voltage V_{REF} is created inside the IC.

The on-time generator circuit can generate the fixed on-time width to control the driver circuit and achieve the voltage regulation if the conventional COT control circuit wants to regulate a high V_{OUT} and an increase in switching loss occurs.

The adaptive on-time control circuits sample the V_{IN} signal and V_{OUT} signal to adjust the on-time width to control the driver circuit and achieve the voltage regulation. The adaptive on-time control circuits for buck converter are as shown Fig. 4.2. S_A and S_B are the switches, L is the output inductor, and R_{CO} is the ESR of the output capacitor C_O. The current source I_{OUT} is the output load current, and R_{D1} and R_{D2} are the feedback resistors to determine the output voltage. V_{FB} is a feedback voltage. The reference voltage V_{REF} is created inside the IC. The adaptive on-time control circuits sample the V_{IN} and V_{OUT} signals to adjust the on-time width to control the driver circuit and achieve voltage regulation. The adaptive on-time control circuits for buck converter are shown Fig. 4.2. The on-time generator circuit samples the V_{IN} and V_{OUT} signals to adjust the on-time width to control the driver circuit and achieve voltage regulation. Two types of adaptive on-time control circuits based on connection form are typically used, namely, constant frequency on-time control circuit and constant current ripple on-time control circuit [18–22].

In adaptive on-time control circuits, the absence of virtual inductor current ripple to add to feedback voltage V_{FB} causes time-delay effects in the loop. Thus, these control circuits suffer from subharmonic oscillations, especially when low-ESR capacitors, such as ceramic capacitors, are used [18–32]. The transfer function of control-to-output $G_{CTO(s)}$ is simplified by Eqs. (4.1)–(4.5) [31, 32].

$$G_{CTO}(s) \approx \frac{1}{1 + s/(\omega_1 \cdot Q_1) + s^2/\omega_1^2} \cdot \frac{(R_{CO} \cdot C_O \cdot s + 1)}{1 + s/(\omega_2 \cdot Q_3) + s^2/\omega_2^2} \tag{4.1}$$

$$\omega_1 = \frac{\pi}{T_{ON}} \tag{4.2}$$

$$Q_1 = \frac{2}{\pi} \tag{4.3}$$

$$\omega_2 = \pi \cdot F_s \tag{4.4}$$

$$Q_3 = \frac{1}{\pi \cdot \left[R_{CO} \cdot C_O - \frac{T_{ON}}{2}\right] \cdot F_s} \tag{4.5}$$

The transfer function indicates that the double pole at half of the switching frequency may move to the right half-plane according to the different capacitor parameters. The critical condition for stability is obtained by Eq. (4.6), which clearly shows the influence of capacitance ripple [18–32].

$$R_{CO} \cdot C_O > \frac{T_{ON}}{2} \tag{4.6}$$

For real capacitors, under the condition 267 kHz of switching frequency, 12 V of input voltage, and 3.3 V of output voltage, the parameter of a single POSCAP capacitor is 330 μF, and its ESR of 4.5 mΩ meets Eq. (4.6). Thus, the system is stable. However, the parameters of the five ceramic capacitors with 22 μF and ESR of 3 mΩ in parallel do not meet Eq. (4.6), so sub-harmonic oscillation occurs.

Experimental results for output voltage and V_{UG} signal with subharmonic oscillation using the ceramic capacitors and without subharmonic oscillation using POSCAP capacitor are as shown in Fig. 4.3. The experimental waveforms include the V_{UG} and V_{OUT} signals, both with and without subharmonic oscillation. The subharmonic oscillation can directly cause very high output voltage ripple with double pulses or multiple pulses. The output voltage with subharmonic oscillation is 284 mV with double pulses higher than that without subharmonic oscillation (67 mV), indicating a difference of about four times ripple magnitude difference in the experimental waveforms. Besides, the switching loss with subharmonic oscillation is larger than that without subharmonic oscillation [18–32].

The subharmonic oscillation can result a large output voltage and the large output voltage may damage the output loading equipment or devices or make the output loading equipment or devices to suffer an error operation. On the other hand, the subharmonic oscillation is also an impact on the load efficiency.

Figure 4.4 shows a comparison of the efficiency of the experimental results with subharmonic oscillation using ceramic capacitors and without subharmonic oscillation using POSCAP capacitor [18–22, 31–35]. The subharmonic oscillation can

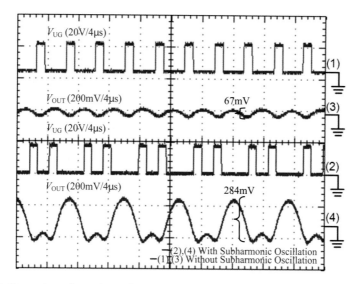

Fig. 4.3 Comparison of experimental results for output voltage and V_{UG} signal with subharmonic oscillation using ceramic capacitors and without subharmonic oscillation using POSCAP capacitor

Fig. 4.4 Comparison of the efficiency of the experimental results with subharmonic oscillation using ceramic capacitors and without subharmonic oscillation using POSCAP capacitor

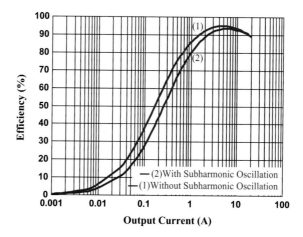

directly cause large switching loss. The difference between the efficiency of the converter that has subharmonic oscillation (43.45%) and the one without subharmonic oscillation (54.4%) is about 11% at a light load of 0.2 A. However, the efficiency of that with subharmonic oscillation is very close to that without subharmonic oscillation at a heavy load of 20 A, because the conduction loss is a major factor that affects the efficiency in this condition.

4.2   Ripple-Based Adaptive On-time Control Circuits with Virtual Inductor Current Ripple for Buck Converters

However, in many applications such as smart phone, netbook, and tablet of consumer products, ceramic capacitors are preferred due to its small size, low output voltage ripple, low cost, and high reliability requirements. One main solution to eliminate subharmonic oscillation is discussed in [18–22, 31–35]. One control circuit of ripple-based adaptive on-time control circuit is by adding a victual inductor current ramp or the internal ramp generator.

Ripple-based adaptive on-time control circuit with virtual inductor current ripple for buck converter is as shown in Fig. 4.5. S_A and S_B are the switches, L is the output inductor, and R_{CO} is the ESR of the output capacitor C_O. The current source I_{OUT} is the output load current, and R_{D1} and R_{D2} are the feedback resistors to determine the output voltage. V_{FB} is a feedback voltage. The reference voltage V_{REF} is created inside the IC. The on-time generator circuit samples the V_{IN} signal and V_{OUT} signal to adjust the on-time width to control the driver circuit and achieve the voltage regulation.

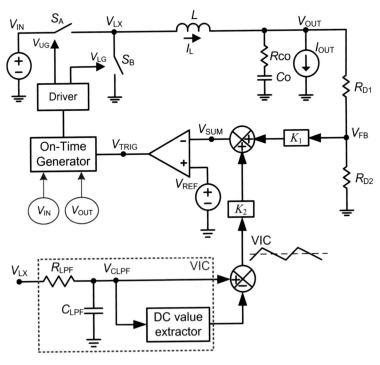

Fig. 4.5 Ripple-based adaptive on-time control circuit with virtual inductor current ripple for buck converter

The use of a virtual inductor current ripple is proposed to alleviate the instability problem because it enhances the effect of the ESR voltage ripple in the feedback voltage. This circuit provides better system stability, especially in all ceramic output capacitors case, which normally has relatively low ESR values. In a virtual inductor current ripple generator, the inductor current ripple waveform can be obtained by integrating the V_{LX} terminal voltage through R_{LPF} and C_{LPF} integrator and removing its DC value by DC value extractor. The DC value of V_{CLPF} is removed by the DC value extractor to generate a virtual inductor current ripple. The virtual inductor current ripple is added in the feedback voltage V_{FB} to enhance the ESR voltage ripple [18–22, 31–35].

The operation waveforms of the virtual inductor current can be explained by the detailed waveforms shown in Fig. 4.6. The rising slope M_{RISE} and falling slope M_{FALL} of V_{CLPF} can be derived by Eqs. (4.7)–(4.8), respectively. The corner frequency of the low-pass filter of R_{LPF} and C_{LPF} just should be designed close to the resonant frequency of the output inductor L and output capacitor C_O, such that the current flowing through R_{LPF} is a nearly constant value to charge or discharge capacitor C_{LPF} during on-time and off-time.

$$M_{RISE} = \frac{V_{IN} - V_{OUT}}{R_{LPF} \times C_{LPF}} \tag{4.7}$$

$$M_{FALL} = \frac{-V_{OUT}}{R_{LPF} \times C_{LPF}} \tag{4.8}$$

It can be seen from Eqs. (4.7)–(4.8) and Fig. 4.7 that virtual ripple V_{CLPF} is proportional to steady-state ESR voltage ripple. The ESR ripple component of the

Fig. 4.6 Operation waveforms of the virtual inductor current

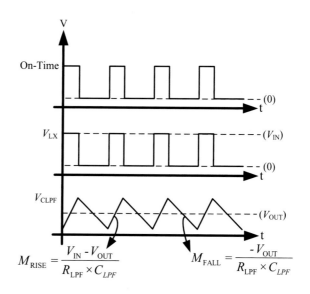

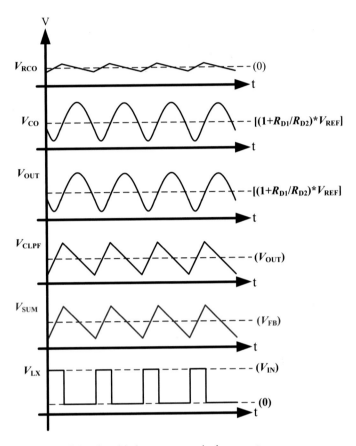

Fig. 4.7 Waveforms of the virtual inductor current ripple generator

negative modulator input V_{SUM} is effectively enhanced by the addition of virtual ripple V_{CLPF} that makes the system more stable [18–22, 31–35].

The virtual inductor current ripple provides good system stability without the sensing current information or adding extra components. If the control circuit can be embedded by a real IC implemented circuit and the signal of the V_{LX} terminal exists in IC, then no extra pin of IC is needed to implement the modified control.

Finally, based on similar operation conditions, a comparison is made between ripple-based adaptive on-time control circuit with virtual inductor current ripple and adaptive on-time control circuit to understand the advantage and superiority of the elimination of subharmonic oscillation. The operation conditions are 267 kHz of switching frequency, 12 V of input voltage, 0.765 V of reference voltage, and 3.3 V of output voltage with no load. The ceramic capacitors are 50 μF, and the ESR of 1 mΩ does not meet Eq. (4.6).

Figure 4.8a shows the simulation results at the steady status for the adaptive on-time control circuit. The red-colored waveform of the output voltage signal is approximately 1.4 V with subharmonic oscillation, and the blue-colored waveform of the V_{RCO} signal is approximately 30 mV of the peak-to-peak voltage. Feedback voltage V_{FB} without the virtual inductor current ripple causes time-delay effects in the loop. Figure 4.8b shows the simulation results at the steady status for the ripple-based adaptive on-time control circuit with virtual inductor current ripple. The red-colored waveform of the output voltage signal is approximately 90 mV without subharmonic oscillation, and the blue-colored waveform of the V_{RCO} signal is approximately 9 mV of the peak-to-peak voltage because the ESR is 1 mΩ. Even the output voltage is without subharmonic oscillation. The virtual inductor current ripple is added to feedback voltage V_{FB}, and the ripple-based adaptive on-time control circuit with virtual inductor current ripple is without time-delay effects in the loop. Thus, the ripple-based adaptive on-time control circuit with virtual inductor current ripple can use ceramic capacitors to reduce the bill of material (BOM) size and achieve a small output voltage ripple.

4.3   Current-Mode Adaptive On-time Control Circuit for Buck Converters

In the adaptive on-time control topology, the absence of inductor current ripple to be added to the capacitor ripple [18–22, 31–35], which is the integration of AC ripples in the inductor current, causes time-delay effects in the loop. Hence, this control scheme suffers from instability caused by subharmonic oscillations because converters with low ESR, such as ceramic capacitors, are used. The current-mode adaptive on-time control circuit for buck converter can be implemented to apply ceramic capacitors with a low ESR because reference voltage V_{REF} and feedback voltage V_{FB} serve as the input terminals of an error amplifier for the current-mode adaptive on-time control circuit and not as the input terminals of a comparator. Thus, the current-mode adaptive on-time control circuit has better noise immunity than the adaptive on-time control circuit.

Figure 4.9 shows the circuit diagram of a current-mode adaptive on-time control circuit for buck converter [36]. In the diagram, S_A and S_B are the switches, L is the output inductor, DCR is the DC resistor of the inductor, R_i is a current sense ratio that is used to transfer an inductor-current signal to a voltage signal V_{CS}, and R_{CO} is the ESR of output capacitor C_O. The diagram shows that the output voltage ripple, which includes inductor-current information, can be directly used as the ramp for modulation. Current source I_{OUT} is the output load current, and R_{D1} and R_{D2} are the feedback resistors that determine the output voltage. The driver circuit uses the input signal on-time width to generate two output signals, V_{UG} and V_{LG}; these two signals should not be turned on at the same time, because such operation causes the system to have a shoot-through problem. The compensation of R_C and C_C should have an optimal design to increase the transient response, if the system simply connects R_C, and it makes the system operate at load line status. Only the feedback

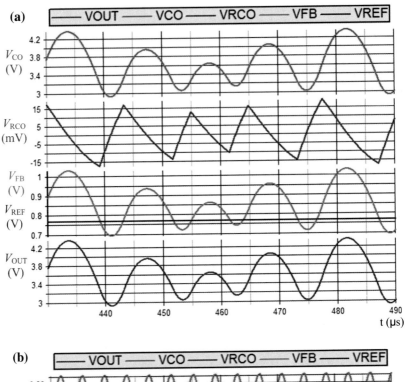

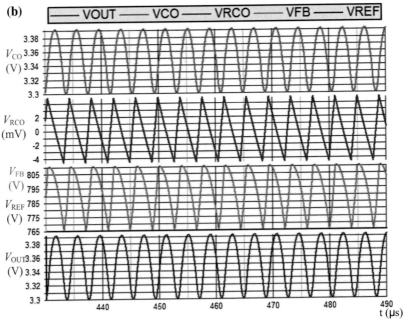

Fig. 4.8 **a** Simulation results at the steady status for adaptive on-time control circuit. **b** Simulation results at the steady status for ripple-based adaptive on-time control circuit with virtual inductor current ripple

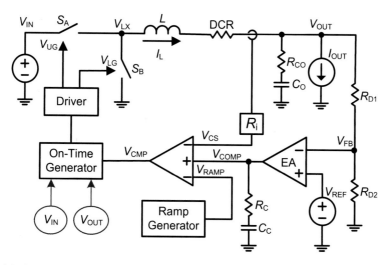

Fig. 4.9 Current-mode adaptive on-time control circuit for buck converter

signal V_{FB} and reference voltage V_{REF} are built inside the IC. The output signal V_{CMP} of the comparator depends on the results on the input signals V_{COMP}, V_{CS}, and V_{RAMP}.

The current-mode adaptive on-time control circuit has an error amplifier, and the output terminal of the error amplifier can amplify a voltage different between V_{FB} and V_{REF}. Thus, it can be used to avoid subharmonic oscillation with double or multiple pulses even when the output capacitor uses ceramic capacitors. The circuit diagram of the current-mode adaptive on-time control circuit is similar to that of the current-mode control circuit. The current-mode adaptive on-time control circuit also needs to sense the inductor current to increase its system loop response. It produces a subharmonic issue when the duty cycle is larger than 50%. A ramp generator is always added to the control loop to prevent the subharmonic issue [36], as shown in Fig. 4.10. The ramp generator is designed according to the inductor current ripple, which is determined by the input voltage, output voltage, and inductance. A general power control IC often embeds a fixed ramp into the control loop, and this ramp can directly increase noise immunity and prevent the system from suffering from a subharmonic issue [36]. Voltage V_A depends on constant voltage V_{CC}, which is also an input signal for VCCS. The output signal of VCCS charges capacitor C to generate fixed ramp V_{RAMP}, which can be calculated using Eq. (4.9).

$$V_{RAMP} = \frac{G_1 \times \left(V_{CC} \times \frac{R_2}{R_1 + R_2}\right)}{C * T_S} \tag{4.9}$$

Under similar operation conditions, a comparison is made between current-mode adaptive on-time control circuit with and without the ramp generator to understand the advantage and superiority of the elimination of the subharmonic issue.

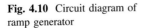

Fig. 4.10 Circuit diagram of ramp generator

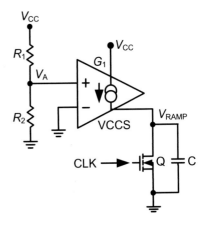

The operation conditions are 395 kHz of switching frequency, 5 V of input voltage, 0.75 V of reference voltage, 3.3 V of output voltage, and 10 A of output load current. The output ceramic capacitor is 44 μF, and its ESR is 1 mΩ.

Figure 4.11a shows the simulation results at the steady status for the current-mode adaptive on-time control circuit without a ramp generator. The red-colored waveform of the output voltage signal is approximately 100 mV with a subharmonic issue, and the blue-colored waveform of the V_{UG} signal is a PWM signal. The V_{UG} signal exhibits a subharmonic issue, so the width of the V_{UG} signal is not the same and the switching frequency cannot be measured, unlike when the V_{UG} signal does not exhibit a subharmonic issue. In addition, the current-mode adaptive on-time control circuit without a ramp generator has a subharmonic issue when the duty cycle is larger than 50%. This subharmonic issue does not result in feedback voltage V_{FB} without the virtual inductor current ripple causing time-delay effects in the loop.

Figure 4.11b shows the simulation results at the steady status for the current-mode adaptive on-time control circuit with a ramp generator. The red-colored waveform of the output voltage signal is approximately 4.6 mV without a sub-harmonic issue, and the blue-colored waveform of the V_{UG} signal is a PWM signal. The V_{UG} signal does not exhibit a subharmonic issue, so the width of the V_{UG} signal is the same, and the switching frequency can be measured at 395.25 kHz.

The current-mode adaptive on-time control circuit has an error amplifier, and the output terminal of the error amplifier can amplify a voltage difference between V_{FB} and V_{REF}. Thus, it can be used to avoid subharmonic oscillation and time-delay effects with double pulses or pulses even when the output capacitor uses ceramic capacitors. The ramp generator is added to voltage V_{CS} when the duty cycle is larger than 50%. The current-mode adaptive on-time control circuit with a ramp generator does not have a subharmonic issue in the loop.

The current-mode adaptive on-time control circuit is highly similar to the current-mode control circuit for buck converter, but the on-time generator of the current-mode adaptive on-time control circuit is significantly different from that of

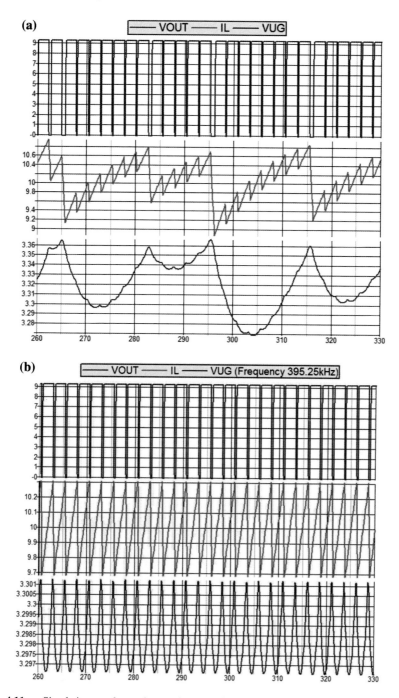

Fig. 4.11 a Simulation results at the steady status for current-mode adaptive on-time control circuit without ramp generator. **b** Simulation results at the steady status for current-mode adaptive on-time control circuit with ramp generator

the current-mode control circuit [36]. The droop can be easily determined based on this difference. The current-mode adaptive on-time control circuit can generate multiple pulses at a variable frequency, and the current-mode control circuit can simply generate a single pulse at a constant frequency when the output load current is changed.

Figure 4.12 shows the simulation results, which can be used to compare the current-mode adaptive on-time control circuit with the current-mode control circuit at the droop [36]. These simulation results use the same operation conditions to compare the current-mode adaptive on-time control circuit with the current-mode control circuit, as shown in Table 4.1 [36]. The current-mode adaptive on-time control circuit can generate multiple pulses with a minimum off-time mechanism to prevent V_{OUT} from decreasing significantly. The current-mode adaptive on-time control circuit is useful for the reduction of V_{OUT} peak-to-peak voltage at the droop. However, the current-mode control circuit can simply increase its PWM pulse width at the same switching frequency at the droop. Comparison between the current-mode adaptive on-time control circuit and the current-mode control circuit at the droop shows that the former can generate more PWM pulses than the latter. Therefore, the current-mode adaptive on-time control circuit can achieve a faster transient response than the current-mode control circuit.

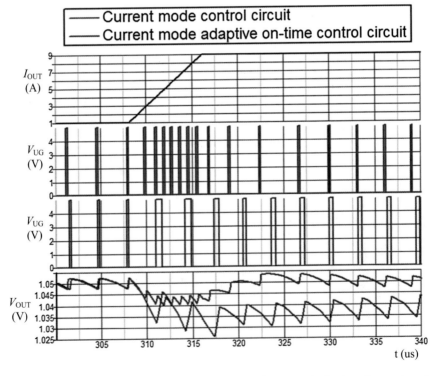

Fig. 4.12 Compare between current-mode adaptive on-time control circuit and current-mode control circuit at the droop

Table 4.1 Operation conditions for current-mode adaptive on-time control and current-mode control circuit for buck converter

V_{IN}	19 V	F_S	320 kHz	L	2.2 µH
V_{OUT}	1.05 V	R_{D1}	8 kΩ	C_{OUT}	330 µF*2
I_{OUT}	1~9 A (8 µs)	R_{D2}	20 kΩ	R_{CO}	5 mΩ
V_{REF}	0.75 V	R_C	1.5 MΩ	C_C	40 pF

4.4 Adaptive On-time Control Circuits with Adaptive Voltage Positioning Design for Voltage Regulators

The AVP design of the adaptive on-time control circuit for VRs is as shown in Fig. 4.13 [3, 14–17, 37–39]. The EA is an error amplifier and V_{COMP} signal is the output of error amplifier. Two input signals of error amplifier are the feedback signal V_{FB} and the V_{VID} voltage, if the user wants to change the output voltage, and the user should be based on VR12 VID code table setting VID bits by I²C interface [40, 41]. S_{11}, S_{12} and S_{13} are the upper-side switches, S_{21}, S_{22} and S_{23} are the lower-side switches, L_1, L_2, and L_3 are the output inductors, and R_{CO} is the ESR of the output capacitor C_O. The current source I_{LOAD} is the output load current. The feedback resistors, R_1 and R_2 can be designed by load line requirement, so the output voltage V_{OUT} is not equal to the V_{VID} voltage. The output voltage V_{OUT} depends on the output load current, if the system is operated from the light load to the heavy load and the output voltage V_{OUT} also should be reduced to meet the load line requirement. The trigger signal V_{TRIG} is the output of comparator to control on-time generator.

The steady state waveforms of the AVP design are as shown in Fig. 4.14. The voltage at the output terminal of the error amplifier is V_{COMP}. The V_{OUT} is larger than V_{OUT_TARGET} in steady state and hence larger the V_{OUT} back to its expected value. The V_{OUT_OFS} voltage is a droop between the V_{OUT} voltage and the V_{OUT_TARGET} voltage. The AVP design of the adaptive on-time control circuit for VRs has a DC offset of the V_{OUT} voltage, however, the VRs should be designed a high accuracy the V_{OUT} voltage, otherwise this offset of the V_{OUT} voltage may impact the load line specification. Refer to the steady-state control signals at the "A" point shown in Fig. 4.14, and Eqs. (4.10)–(4.16) are obtained [3, 14–17, 37–39].

$$V_{CS,Vally} = V_{COMP,Peak} \tag{4.10}$$

$$V_{VID} + R_i \cdot (I_L - \frac{\Delta I_L}{2}) = V_{COMP} \tag{4.11}$$

$$V_{VID} + R_i \cdot (I_L - \frac{\Delta I_L}{2}) = V_{VID} + A_V \cdot (V_{VID} - (V_{OUT} - \frac{\Delta V_{OUT}}{2})) \tag{4.12}$$

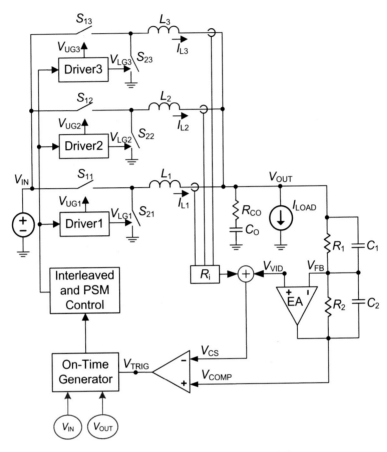

Fig. 4.13 AVP design of the adaptive on-time control circuit for VRs

$$A_V \cdot V_{OUT} = A_V \cdot \left(V_{VID} + \frac{\Delta V_{OUT}}{2}\right) - R_i \cdot \left(I_L - \frac{\Delta I_L}{2}\right) \tag{4.13}$$

$$V_{OUT} = \left(V_{VID} + \frac{\Delta V_{OUT}}{2}\right) - \frac{R_i}{A_V} \cdot \left(I_L - \frac{\Delta I_L}{2}\right) \tag{4.14}$$

$$R_{DROOP} = \frac{R_i}{A_V} \tag{4.15}$$

$$A_V = \frac{R_2}{R_1} \tag{4.16}$$

A_V is the desired error amplifier gain. R_i is the internal current sense amplifier gain. R_{DROOP} is the current sense resistor. This control also implements the AVP

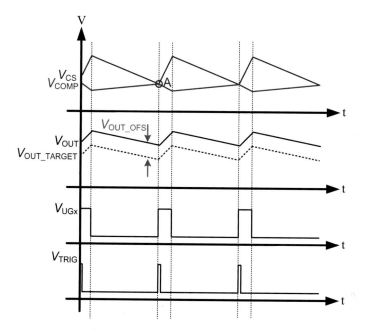

Fig. 4.14 Steady state waveforms of the AVP design

function easily. R_{DROOP} should be designed to determine the load line. It is the equivalent load line resistance as well as the desired static output impedance.

An optimized compensation of a multiphase voltage regulator allows for the best possible load step response of its output. A type-II compensator with one pole and one zero is adequate for proper compensation. A prior design procedure shows how to decide the resistive feedback components of an error amplifier gain. C_1 and C_2 can be calculated for compensation by Eq. (4.17) [40, 41]. The target is to achieve constant resistive output impedance over the widest possible frequency range.

$$G_{\text{VC}}(s) = \frac{R_{\text{i}}}{R_{\text{DROOP}}} \frac{1 + s/(R_1 \times C_1)}{1 + s/(R_2 \times C_2)} \tag{4.17}$$

4.5   Summary

The conventional COT control circuit is more suitable to be used for fast load transient power supplies application and it can generate the fixed on-time width to control the driver circuit and achieve the voltage regulation. The on-time generator circuit of the conventional COT control circuit can generate the fixed on-time width to control the driver circuit and achieve the voltage regulation if the conventional

COT control circuit wants to regulate a high V_{OUT} and an increase in switching loss occurs.

The adaptive on-time control circuits sample the V_{IN} signal and V_{OUT} signal to adjust the on-time width to control the driver circuit and achieve the voltage regulation. Two types of the adaptive on-time control circuits based on connection form are typically used, namely, constant frequency on-time control circuit and constant current ripple on-time control circuit.

References

1. E. Stanford, Power technology roadmap for microprocessor voltage regulators. Presentation at PSMA, Feb 2003
2. 2001 technology roadmap for semiconductors. Available: http:// www.intel.com/research/silicon/AlanAllanIEEEComputer0102.pdf
3. K. Yao, Y. Ren, J. Sun, K. Lee, M. Xu, J. Zhou, F.C. Lee, Adaptive voltage position design for voltage regulators, in *Proceedings IEEE Applied Power Electronics Conference and Exposition conference*, pp. 272–278, 2004
4. H. Mao, L. Yao, C. Wang, I. Batarseh, Analysis of Inductor current sharing in non isolated and isolated Multiphase dc–dc converters. IEEE Trans. Ind. Electron. **54**(6), 3379–3388, Dec 2007
5. P.-W. Lee, Y.-S. Lee, D.K.W. Cheng, X.-C. Liu, Steady-state analysis of an interleaved boost converter with coupled inductors. IEEE Trans. Ind. Electron. **47**(4), 787–795 (2000)
6. H.N. Nagaraja, D. Kastha, A. Patra, Design principles of a symmetrically coupled inductor structure for multiphase synchronous buck converters. IEEE Trans. Ind. Electron. **58**(3), 988–997 (2011)
7. L.-P. Wong, D.K.-W. Cheng, M.H.L. Chow, Y.-S. Lee, Interleaved three-phase forward converter using integrated transformer. IEEE Trans. Ind. Electron. **52**(5), 1246–1260 (2005)
8. J. Abu-Qahouq, H. Mao, I. Batarseh, Multiphase voltage-mode hysteretic controlled dc–dc converter with novel current sharing. IEEE Trans. Power Electron. **19**(6), 1397–1407 (2004)
9. H. Mao, L. Yao, C. Wang, I. Batarseh, Analysis of inductor current sharing in non isolated and isolated multiphase dc–dc converters. IEEE Trans. Ind. Electron. **54**(6), 3379–3388, Dec 2007
10. P.-W. Lee, Y.-S. Lee, D.K.W. Cheng, X.-C. Liu, Steady-state analysis of an interleaved boost converter with coupled inductors. IEEE Trans. Ind. Electron **47**(4), 787–795 (2000)
11. H.N. Nagaraja, D. Kastha, A. Patra, Design principles of a symmetrically coupled inductor structure for multiphase synchronous buck converters. IEEE Trans. Ind. Electron **58**(3), 988–997 (2011)
12. L.-P. Wong, D.K.-W. Cheng, M.H.L. Chow, Y.-S. Lee, Interleaved three-phase forward converter using integrated transformer. IEEE Trans. Ind. Electron. **52**(5), 1246–1260 (2005)
13. J. Abu-Qahouq, H. Mao, I. Batarseh, Multiphase voltage-mode hysteretic controlled dc–dc converter with novel current sharing. IEEE Trans. Power Electron. **19**(6), 1397–1407 (2004)
14. J.R. Huang, C.H. Wang, C.J. Lee, K.L. Tseng, and D. Chen, Native AVP control method for constant output impedance of DC power converters, in *Proceedings IEEE Power Electronics Specialists Conference*, pp. 2023–2028, 2007
15. A. Waizman, C.Y. Chung, Resonant free power network design using extended adaptive voltage positioning (EAVP) methodology. IEEE Trans. Adv. Packag. **24**, 236–244 (2001)
16. M. Lee, D. Chen, K. Huang, E. Tseng and B. Tai, Compensator design for adaptive voltage position (AVP) for multiphase VRMs, in *Proceedings IEEE Power Electronics Specialists Conference*, 2006

17. K. Yao, Y. Meng, P. Xu, F.C. Lee, Design considerations for VRM transient response based on the output impedance, in *Proceedings IEEE Applied Power Electronics Conference and Exposition conference*, pp. 14–20, 2002
18. W.W. Chen, J.F. Chen, T.J. Liang, L.C. Wei, J.R. Huang, W.Y. Ting, A novel quick response of RBCOT with VIC ripple for buck converter. IEEE Trans. Power Electron. **28**, 4299–4308 (2013)
19. W.W. Chen, J.F. Chen, T.J. Liang, J.R. Huang, L.C. Wei, W.Y. Ting, Implementing dynamic quick response with high-frequency feedback control of the deformable constant on-time control for buck converter on-chip. IET Power Electron **6**(4), 383–391 (2013)
20. W.W. Chen, J.F. Chen, T.J. Liang, S.F. Hsiao, J.R. Huang, W.Y. Ting, Improved transient response using HFFC circuit of the CCRCOT with native AVP design for voltage regulators. IET Power Electron **6**, 1948–1955 (2013)
21. W.W. Chen, J.F. Chen, T.J. Liang, J.R. Huang, and W.Y. Ting, Improved Transient Response Using HFFC in Current-Mode CFCOT Control for Buck Converter, in *Proceedings IEEE International Conference on Power Electronics and Drive Systems (PEDS)*, pp. 546–549
22. C.J. Chen, D. Chen, C.W. Tseng, C.T. Tseng, Y.W. Chang, K.C. Wang, A Novel Ripple-Based Constant On-time Control with Virtual Inductor Current Ripple for Buck Converter with Ceramic Output Capacitors, in *Proceedings IEEE Applied Energy Conversion Congress and Exposition conference*, pp. 1244–1250, 2011
23. R. Redl, J. Sun, Ripple-based control of switching regulators an overview. IEEE Trans. Power Electron. **24**, 2669–2680 (2009)
24. J. Sun, Characterization and performance comparison of ripple-based control for voltage regulator modules. IEEE Trans. Power Electron. **21**, 346–353 (2006)
25. W. Huang, A new control for multi-phase buck converter with fast transient response, in *Proceedings IEEE Applied Power Electronics Conference and Exposition conference*, pp. 273–279, 2001
26. J. Li, F.C. Lee, Modeling of V2 current-mode control, in *Proceedings IEEE Applied Power Electronics Conference and Exposition conference*, pp. 298–304, 2009
27. K.D.T. Ngo, S.K. Mishra, M. Walters, Synthetic-ripple modulator for synchronous buck converter. Proc. IEEE Power Electron. **3**, 148–151 (2005)
28. Y.H. Lee, S.J. Wang, K.H. Chen, Quadratic differential and integration technique in V2 control buck converter with small ESR capacitor. Proc. IEEE Trans. Power Electron. **25**, 829–838 (2010)
29. M.Y. Yen, P. Mok, A constant frequency output-ripple voltage-based buck converter without using large ESR capacitor. IEEE Trans. Circuits Syst. **55**, 748–752 (2008)
30. K.Y. Cheng, F. Yu, P. Mattavelli, F.C. Lee, Characterization and performance comparison of digital V2-type constant on-time control for buck converters, *IEEE Control and Modeling for Power Electronics conference*, pp. 1–6, June 2010
31. J. Li, Current-mode control: modeling and its digital application Ph.D. thesis, Virginia Polytechnic Institute and State University, 2009
32. J. Li and F.C. Lee, New modeling approach and equivalent circuit representation for current-mode control. IEEE Trans. Power Electron. 1218–1230, May 2010
33. S.J. Wang, Y.H. Lee, Y.C. Lai, K.H. Chen, Quadratic differential and integration technique in V2 control buck converter with small ESR capacitor, in *Proceedings IEEE Custom Integrated Circuits Conference*, pp. 211–214, 2009
34. R. Redl, G. Reizik, Switched noise filter for the buck converter using the output ripple as the PWM ramp, in *Proceedings IEEE Appl. Power Electron. Conference*, pp. 918–924, 2005
35. R. Redl and T. Schiff, A new family of enhanced ripple regulators for power-management applications, in *Proceedings of International Exhibition and Conference Eur.*, Nuremburg, Germany, pp. 255–268, 2008
36. W.W. Chen, J.F. Chen, T.J. Liang, Dynamic ramp control in current-mode adaptive on-time control for buck converter on chip, in *Proceedings of IEEE Future Energy Electronics Conference and ECCE Asia (IFEEC 2017–ECCE Asia)*, pp. 280–285, 2017

37. P.L. Wong, Performance improvements of multi-channel interleaving voltage regulator modules with integrated coupling inductors, Dissertation of Virginia Polytechnic Institute and State University, March 2001
38. S.K. Mishra, Design-oriented analysis of modern active droop controlled power supplies. IEEE Trans. Ind. Electron. **56**(9), 3704–3708 (2009)
39. J.A.A. Qahouq, V. Arikatla, Power converter with digital sensorless adaptive voltage positioning control scheme. IEEE Trans. Ind. Electron. **58**(9), 4105–4116 (2010)
40. Richtek Tech. Corp., Dual output 3-Phase + 2-Phase PWM controller for CPU and GPU Core power supply, RT8885A Datasheet, 2012
41. Richtek Tech. Corp., Multi-Phase PWM controller for CPU core power supply, RT8859 M Datasheet, 2014

Chapter 5
Adaptive On-time Control Circuit for Buck Converters

5.1 Increasing Light Load Efficiency with PSM Mode

Today's products, particularly those designed to reduce standby power loss and increase system loop response, feature high performance and low loss because of their system design. The conventional fixed-frequency PWM control scheme for power converters [1–13] is commonly used with current-mode control instead of voltage-mode control. However, adaptive on-time control circuits [14–25] can achieve a faster transient response than the PWM control scheme because the former uses a comparator to control the on-time generator without an error amplifier, as shown in Fig. 5.1. Thus, adaptive on-time control circuits do not exhibit system loop delay from the error amplifier. However, adaptive on-time control circuits do not meet the requirements of output equipment or devices to achieve a faster transient response. This chapter shows that adaptive on-time control circuits with a quick dynamic response can achieve a faster transient response than those without a quick dynamic response for buck converters.

In Fig. 5.1, S_A and S_B are the switches, L is the output inductor, and R_{CO} is the ESR of the output capacitor C_O. The current source I_{OUT} is the output load current, and R_{D1} and R_{D2} are the feedback resistors to determine the output voltage. V_{FB} is a feedback voltage. The reference voltage V_{REF} is created inside the IC. The on-time generator circuit samples the V_{IN} signal and V_{OUT} signal to adjust the on-time width to control the driver circuit and achieve the voltage regulation.

The ZCD function [26–30] is widely used to maintain the inductor current at zero under light load for buck converter ICs. Thus, the ZCD function can prevent reverse current from flowing back to the source to improve conversion efficiency. Under light load, the ZCD function for each switching period can be divided into three operation modes. In mode 1, the S_A switch is turned ON, and buck converter begins to charge the output capacitors and delivers energy to output equipment or devices. In mode 2, the S_A switch is turned OFF immediately after the S_B switch is turned ON to discharge from the output capacitors and deliver energy to output

© Springer Nature Singapore Pte Ltd. 2018
W.-W. Chen and J.-F. Chen, *Control Techniques for Power Converters with Integrated Circuit*, Power Systems,
https://doi.org/10.1007/978-981-10-7004-4_5

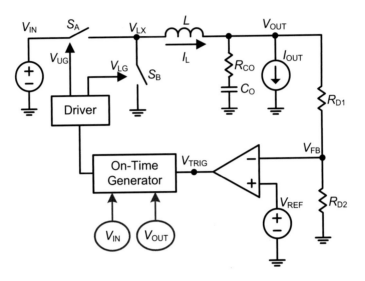

Fig. 5.1 Adaptive on-time control circuits for buck converter

equipment or devices. In mode 3, the inductor current is lower than or equal to zero, and the S_B switch is turned OFF. The inductor current flows through the body diode of the S_B switch until the next duty cycle is needed to drive the S_A switch. Thus, switching loss can be reduced significantly, as shown in Fig. 5.2. Figure 5.2 shows the V_{LX} voltage and the I_L current. V_{LX} can obtain the control signal information of

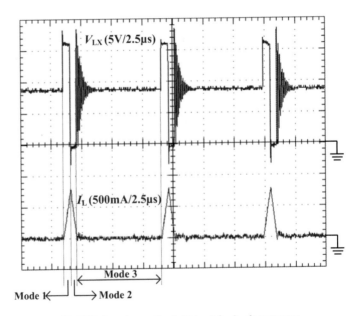

Fig. 5.2 Operation with ZCD function at the light load for buck converter

V_{UG} and V_{LG} voltages. When the V_{LX} voltage is at the maximum level, it is equal to the input voltage because the S_A switch is turned ON at this status. When the V_{LX} voltage is at a negative level, it is equal to the voltage drop of R_{DS} for the S_B switch because the S_B switch is turned ON at this status. The S_A and S_B switches are turned OFF when the I_L current is equal to zero, and the V_{LX} voltage is equal to the V_{OUT} voltage. To save cost, most IC products include S_A and S_B switches on chip, so the user can measure the V_{LX} voltage to obtain V_{UG} and V_{LG} voltages.

Pulse-skipping modulation (PSM) mode [26–30] is also widely used in ICs to increase the efficiency under light load for buck converters. The PSM mode can reduce the number of control pulses under light loads. The conventional fixed-frequency PWM control scheme needs an extra circuit to skip the original control pulses to operate at PSM mode. However, adaptive on-time control circuits can directly achieve a native PSM mode at a light load because the control pulses of these circuits depend on the output loading condition. Thus, the number of control pulses at a light load is less than that at a heavy load. The usual PSM mode of combining adaptive on-time control circuits with the ZCD function can best reduce the switching loss.

To further understand the advantage and superiority of this PSM mode, the plans based the same operation conditions to compare the adaptive on-time control circuits with PSM mode and the current-mode control circuit at the light load with the ceramic output capacitors 44 μF.

The conditions are as follows:

(1) Input voltage (V_{IN}): 12 V
(2) Output voltage (V_{OUT}): 1.05 V
(3) Output load current (I_{OUT}): 10 mA
(4) Switching frequency (F_S): 630 kHz
(5) MOSFET (S_A, S_B): BSC0909NS * 3
(6) Feedback resistors (R_{D1}, R_{D2}): 13.2 kΩ and 12 kΩ
(7) Main inductor (L): IHLP4040DZER1R0MA1 (1 μH)
(8) Output capacitors (C_O): 22 μF/6.3 V (R_{CO}: 3 mΩ) * 2
(9) Reference voltage (V_{REF}): 0.5 V.

Figure 5.3 compares the experimental results of the adaptive on-time control circuits with PSM mode and current-mode control circuit at the light load. Figure 5.3a represents the V_{LX} voltage and the I_L current for current-mode control circuit at the light load. Figure 5.3b represents the V_{LX} voltage and the I_L current for adaptive on-time control circuits with PSM mode at the light load. The signals are measured based on the same output load current. The switching frequency of the current-mode control circuit maintains 630 kHz. The switching frequency of the adaptive on-time control circuits with PSM mode is based on the output loading condition and the switching frequency is measured 33.3 kHz at output current 10 mA. The number of control pulses for current-mode control circuit is larger than adaptive on-time control circuits with PSM mode, so the current-mode control circuit has a large switching loss. In addition, the I_L current of the adaptive on-time

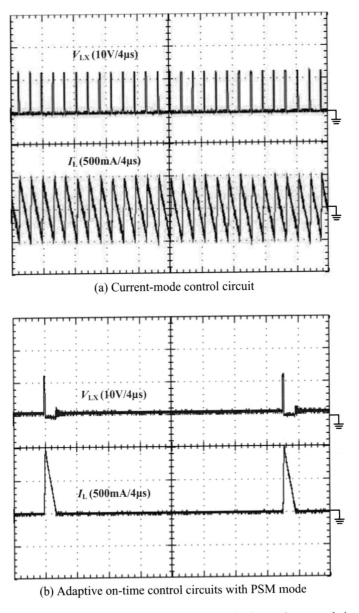

(a) Current-mode control circuit

(b) Adaptive on-time control circuits with PSM mode

Fig. 5.3 Comparison of the experimental results between adaptive on-time control circuits with PSM mode and current-mode control circuit at the light load

control circuits with PSM mode is maintained zero current during 28 μs, so it is useful to save conduction loss on the R_{DS} for the S_B switch, because the inductor current is through the body diode of the S_B switch. Based on these reasons, the

adaptive on-time control circuits with PSM mode can significantly achieve a better
efficiency at the light load.

Figure 5.4 shows one control pulse of the experimental results of adaptive
on-time control circuits with PSM mode and the current-mode control circuit at a

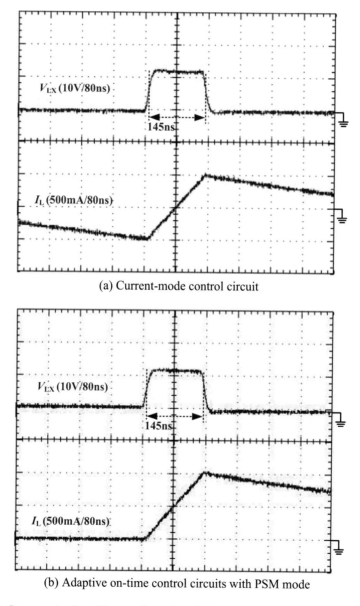

(a) Current-mode control circuit

(b) Adaptive on-time control circuits with PSM mode

Fig. 5.4 One control pulse of the experimental results between adaptive on-time control circuits
with PSM mode and current-mode control circuit at the light load

Fig. 5.5 Comparison of the efficiency results between adaptive on-time control circuits with PSM mode and current-mode control circuit

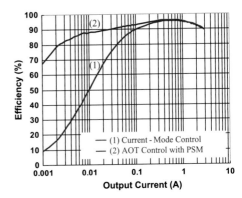

light load. The signals are measured based on the same output load current. The width of the control pulse of the current-mode control circuit is 145 ns, and it is equal to the width of the control pulse of the adaptive on-time control circuits with PSM mode. The adaptive on-time control circuits can sample the V_{IN} and V_{OUT} voltages to control the on-time, similar to the duty cycle of the current-mode control circuit. Even when the V_{IN} and V_{OUT} voltages are changed, the system still maintains the same duty cycle. The same width of the control pulse can generate a similar rise time of the I_L current. In addition, the peak-to-peak I_L current of the current-mode control circuit is 1 A, and it is equal to the width of the control pulse of the adaptive on-time control circuits with PSM mode.

Figure 5.5 compares the efficiency results between adaptive on-time control circuits with PSM mode and current-mode control circuit. Adaptive on-time control circuits with PSM mode at the light load results in lower switching frequency and lower switching loss compared with current-mode control circuit. When the output load current is <0.2 A, the efficiency with adaptive on-time control circuits with PSM mode is significantly improved. Because the current-mode control circuit at the light load, it still allows this converter to remain at a fixed frequency, and the duty cycle becomes the input voltage and the output voltage of the decision. This fixed frequency results in a large switching loss, and the converter can no longer achieve a low loss. Minimizing the standby power loss is usually preferred to reducing switching loss.

5.2   On-time Generator Circuit of the Adaptive On-time Control Circuits for Buck Converters

Two types on-time generator circuit of the adaptive on-time control circuits based on connection form are typically used, namely, constant frequency on-time control circuit and constant current ripple on-time control circuit [1, 31–34].

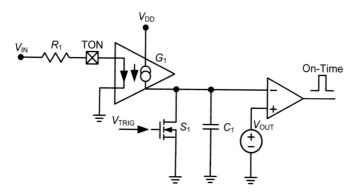

Fig. 5.6 On-time generator circuit of the constant frequency on-time control circuit

On-time generator circuit of the constant frequency on-time control circuit is as shown in Fig. 5.6. The TON pin is inside the IC, and the resistor R_1 can be placed between TON pin and input voltage to determine the on-time width. V_{OUT} needs to be monitored to limit the voltage from capacitor C_1 by Eq. (5.1). V_{OUT} divided by V_{IN} is the duty cycle, and R_1, C_1, and G_1 are constant values, like the switching period T_S, as shown in Eq. (5.2). G_1 is the gain of the current-controlled current source. Even if the input voltage V_{IN} and output voltage V_{OUT} are changed, the system still maintains the same switching frequency [1, 31–34].

$$T_{ON} = \frac{V_{OUT}}{V_{IN}} \times \frac{R_1 \times C_1}{G_1} \qquad (5.1)$$

$$T_{ON} = \frac{V_{OUT}}{V_{IN}} \times T_S \qquad (5.2)$$

Constant frequency on-time control circuit is different from the conventional COT control circuit because it prevents the generation of the same on-time width at a high V_{OUT}. Regardless of changes in V_{IN} or V_{OUT}, the conventional COT control circuit still generates the same on-time width. If the conventional COT control circuit wants to regulate a high V_{OUT} and needs to generate more on-time pulses, an increase in switching loss occurs. Thus, constant frequency on-time control circuit is suitable for a wide range output voltage.

Experimental results of the switching frequency versus the input voltage for constant frequency on-time control circuit are as shown in Fig. 5.7. The operation conditions are set at 3.3 V output voltage and 0.1 A output current to measure the switching frequency for 4–21 V input voltage. When R_1 is equal to 250 kΩ, connecting V_{IN} and TON pin, the switching frequency is 267 kHz, regardless of changes in V_{IN}. If R_1 is reduced to 150 kΩ, the switching frequency changes from 267 to 444 kHz, because R_1 directly affects the on-time width. R_1 and the switching frequency are inversely related.

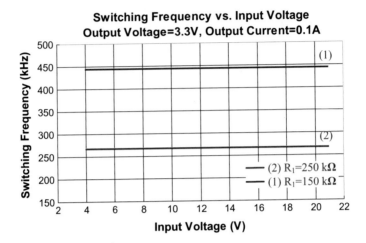

Fig. 5.7 Experimental results of the switching frequency versus the input voltage for constant frequency on-time control circuit

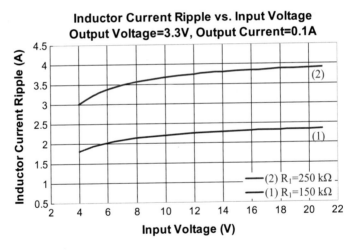

Fig. 5.8 Experimental results of the inductor current ripple versus the input voltage for constant frequency on-time control circuit

Experimental results of the inductor current ripple versus the input voltage for constant frequency on-time control circuit are as shown in Fig. 5.8. The operating conditions are set 3.3 V output voltage and 0.1 A output current to measure the inductor current ripple for 4–21 V input voltage. If the inductance is invariant, then the inductor current ripple is related to the voltage difference between the input voltage and output voltage and is also dependent on the on-time T_{ON}, as shown by Eq. (5.3). By substituting Eq. (5.2) into Eq. (5.3), the inductor current ripple is obtained as Eq. (5.4). The inductor current ripple increases as the input voltage is increased. Hence, it does not maintain a constant value. When R_1 is increased from 150 to 250 kΩ, the on-time width is also increased, resulting in a larger inductor current ripple.

$$\Delta I_L = \frac{V_{IN} - V_{OUT}}{L} \times T_{ON} \tag{5.3}$$

$$\Delta I_L = \frac{V_{OUT}(1 - (V_{OUT}/V_{IN}))}{L} \times T_S \tag{5.4}$$

On-time generator circuit of the constant current ripple on-time control circuit is as shown in Fig. 5.9. The input voltage and output voltage are sampled to determine the on-time width. The TON is an IC pin connected to V_{IN} through a resistor. This circuit also samples V_{OUT} by Eq. (5.5). The voltage drop between the input voltage and output voltage depends on the inductor current ripple, as shown by Eq. (5.6). Therefore, even if the voltage drop between the input voltage and output voltage is changed, the inductor current ripple maintains a constant value. Hence, the constant current ripple on-time control circuit is suitable for applications with different the input voltage and output voltage, requiring the same output voltage ripple, such as CPUs [1, 31–34].

$$T_{ON} = \frac{R_1}{V_{IN} - V_{OUT}} \times C_1 \times V_{REF1} \tag{5.5}$$

$$T_{ON} = \frac{L}{V_{IN} - V_{OUT}} \times \Delta I_L \tag{5.6}$$

Experimental results of the switching frequency versus the input voltage for constant current ripple on-time control circuit are as shown in Fig. 5.10. The operation conditions are set at 3.3 V output voltage and 0.1 A output current to

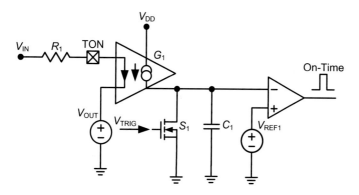

Fig. 5.9 On-time generator circuit of the constant current ripple on-time control circuit

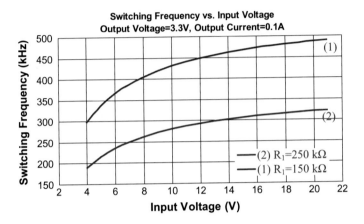

Fig. 5.10 Experimental results of the switching frequency versus the input voltage for constant current ripple on-time control circuit

measure the switching frequency for 4–21 V input voltage. When R_1 is equal to 150 kΩ, connecting V_{IN} and TON pin, the switching frequency increases from 296 to 489 kHz when the input voltage is increased from 4 to 21 V. Increasing R_1 from 150 to 250 kΩ, the switching frequency increases from 188 to 322 kHz when the input voltage is increased from 4 to 21 V.

Experimental results of the inductor current ripple versus the input voltage for constant current ripple on-time control circuit are as shown in Fig. 5.11. The operation conditions are set at 3.3 V output voltage and 0.1 A output current to measure the inductor current ripple for 4–21 V input voltage. If the inductance is

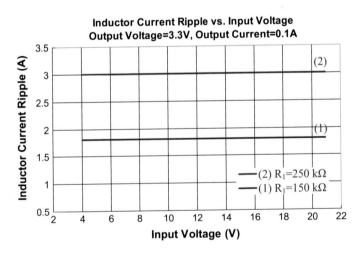

Fig. 5.11 Experimental results of the inductor current ripple versus the input voltage for constant current ripple on-time control circuit

invariant, then the inductor current ripple is related to the voltage drop between the input voltage and output voltage and is also dependent on the on-time T_{ON}. The inductor current ripple maintains a consistent value. When R_1 is increased from 150 to 250 kΩ, the inductor current ripple also increases from 1.8 to 3.0 A.

5.3 Comparison of Quick Dynamic Response and Conventional Quick Response of the On-time Generator Circuit

Two types of improved transient response circuits are typically used, namely, external setting by component or voltage source and internal fixed by IC design. Figure 5.12 shows the external setting quick response of the on-time generator circuit for constant frequency on-time control circuit. The operation principle of the external setting quick response requires the addition of a voltage source V_{QRSET} to determine the QR-time width. V_{QRSET} is connected to the QRSET pin. If the QR-time needs a long QR-time width, V_{QRSET} should be larger than V_{OUT} [1, 31–34].

V'_{OUT} determines the peak voltage of C_1. Thus, if V'_{OUT} is high, the peak voltage increases, causing a long on-time width. V'_{OUT} is implemented to provide a ripple-less voltage because it can keep the switching frequency constant without

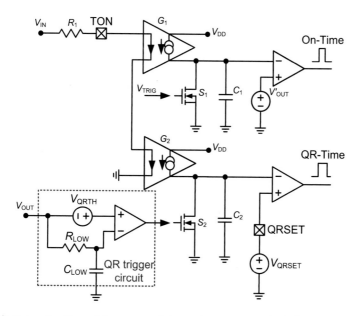

Fig. 5.12 External setting quick response of the on-time generator circuit for constant frequency on-time control circuit

under output voltage ripple and noise interference. V'_{OUT} is implemented to sample the Phase signal through a second-order low-pass filter, producing a similar V_{OUT} signal. Therefore, V'_{OUT} is ripple-less and has slower transient response than the V_{OUT} signal.

The QR trigger circuit samples V_{OUT} and uses the low-pass filter to generate V_{OUT} delay. This trigger circuit also needs the voltage signal V_{QRTH}. In the steady-state operation, the V_{OUT} drop cannot trigger the QR-time. When this abrupt voltage drop is lower than V_{QRTH}, the output of QR trigger circuit changes from a high-level signal to a low-level signal to turn OFF the S_2 switch and then the on-time generator circuit generates the QR-time width. The low-pass filter frequency for R_{LOW} and C_{LOW} should be designed much smaller than the switching frequency which can avoid this system to fail in its operation with QR-time. V_{QRTH} can be designed in the IC or set by the user through the QRTH pin. The main advantage of the external setting quick response is that it can depend on the load conditions to design the V_{QRSET} value, generating a suitable QR-time width in this system. However, this method requires two extra IC pins to achieve the quick response function [1, 31–34].

Internal fixed quick response of the on-time generator circuit for constant current ripple on-time control circuit is as shown in Fig. 5.13. This circuit has two on-time generator circuits to generate the on-time width and QR-time width. When the constant current ripple on-time control circuit is operated at steady state, the system generates the on-time control signal to drive the switches, thereby regulating the output voltage. The output load current is changed from a light load to a heavy load, which makes the output voltage dropped significantly thereby causing the system to generate the QR-time control signal and achieve the quick response function.

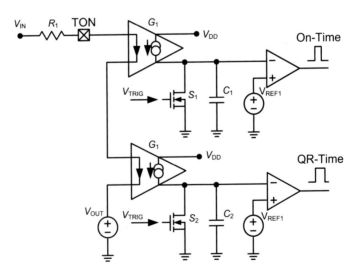

Fig. 5.13 Internal fixed quick response of the on-time generator circuit for constant current ripple on-time control circuit

The QR-time width must be longer than the on-time width. The implementation of the QR-time generator circuit is similar to that of the on-time generator circuit. The only difference is that C_2 must be greater than C_1 [1, 31–34].

Implementing the internal fixed quick response can also improve the transient response for constant current ripple on-time control circuit. The main advantage of this circuit is that it does not need an extra IC pin to achieve the quick response function. However, it has some disadvantages: (1) The IC cannot adjust the QR-time width. Hence, when the output load current is changed from a light load to a heavy load, the system still outputs the same width of the QR-time control signal; (2) Designing a hysteresis level to generate the QR-time control signal is difficult. Hence, when the output load current is different, this circuit may cause the system to fail in its operation with QR-time control signal.

Quick dynamic response of the on-time generator circuit for constant frequency on-time control circuit is proposed in this chapter, as shown in Fig. 5.14. This quick dynamic response does not require an extra pin to achieve the quick response function of the IC. This quick dynamic response consists of a resistor R_q and a capacitor C_q connected in series between V_{OUT} and TON pin. It operates under the principle that the high-pass filter allows only high frequency signal to pass from V_{OUT} to the TON pin, as shown in Eq. (5.7). However, the frequency F_{RC} of the high-pass filter must be larger than or equal to the switching frequency F_S, so that the steady state operation of the system is not affected. Finally, the user may design R_q and C_q based on the worst-case operation (maximum output loading step) [1, 31–34].

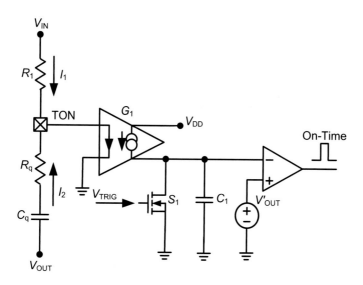

Fig. 5.14 Quick dynamic response of the on-time generator circuit for constant frequency on-time control circuit

$$F_{RC} = \frac{1}{2\pi \times R_q \times C_q} \geq F_S \qquad (5.7)$$

V'_{OUT} determines the peak voltage of C_1. Thus, if V'_{OUT} is high, the peak voltage increases, causing a long on-time width. V'_{OUT} is implemented to provide a ripple-less voltage because it can keep the switching frequency constant without under output voltage ripple and noise interference. V'_{OUT} is implemented to sample the Phase signal through a second-order low-pass filter producing a similar V_{OUT} signal. Therefore, V'_{OUT} is ripple-less and has a slower transient response than V_{OUT}.

When a light load quickly transforms to a heavy load, V_{OUT} momentarily drops through the coupling by C_q, which inducing R_q to cause a voltage drop. The voltage drop of R_q will form a current of I_2. Thus, R_q needs to be designed first because I_2 can directly change the on-time width as shown in Fig. 5.15.

However, a longer on-time width induces the converter to deliver more energy from the input terminal to the output loading. If R_q is reduced, the on-time width becomes longer. C_q should be designed with the frequency of high-pass filter. The calculation of the Laplace transform formula is shown in Eq. (5.8).

$$T_{ON}(s) = \frac{C_1}{G_1\big(((V_{IN})/(R_1)) + ((sC_qV_{OUT})/(1 + sC_qR_q))\big)} \times V'_{OUT} \qquad (5.8)$$

The advantages of this quick dynamic response of the on-time generator circuit are as follows: (1) It clearly generates a longer on-time width that is proportional to the output voltage drop; (2) Adaption to the on-time width depends on the load conditions of the system; (3) It only uses one on-time generator circuit; thus, it is very convenient to design and apply; (4) It does not require an extra pin to achieve the quick response function.

Fig. 5.15 Control signals for quick dynamic response of the on-time generator circuit for constant frequency on-time control circuit

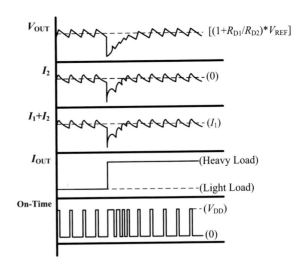

5.4 Experimental Results

The verification of the experimental and simulation results is conducted to prove the feasibility and performance with quick dynamic response of the constant frequency on-time control circuit for Buck converter.

The specifications are as follows:

(1) Input voltage (V_{IN}): 12 V
(2) Output voltage (V_{OUT}): 3.3 V
(3) Output load current (I_{OUT}): 18 A @ 1.2857 A/µs
(4) Switching frequency (F_S): 267 kHz
(5) MOSFET (S_A, S_B): BSC0909NS * 3
(6) Feedback resistors (R_{D1}, R_{D2}): 47 and 15 kΩ
(7) Main inductor (L): IHLP4040DZER1R0MA1 (1 µH)
(8) Output capacitors (C_O): 330µF/6.3 V (R_{CO}:4.5 mΩ) * 1
(9) Reference voltage (V_{REF}): 0.8 V
(10) Quick dynamic response: G_1 = 1(A/A); R_1 = 250 kΩ; C_1 = 15 pF; R_q = 500 Ω; C_q = 1.2 nF; F_{RC} = 262 kHz
11) External setting quick response: R_{LOW} = 600 kΩ; C_{LOW} = 2 pF; V_{QRSET} = 4.125 V; V_{QRTH} = 100 mV

Figure 5.16 shows the chip layout of the control IC. The driver circuit and TON control are both occupied, marking an area with a die size of 880 µm × 880 µm.

Evaluation board of the constant frequency on-time control circuit with quick dynamic response for Buck converter is as shown in Fig. 5.17.

SIMPLIS simulation results at the load transient with quick dynamic response (see Fig. 5.14) and without quick response (see Fig. 5.6) are as shown in Fig. 5.18. Without quick response cannot directly change its on-time width at the load transient. The blue-colored waveform represents the output voltage signal without quick response, and the red-colored waveform represents the output voltage signal with quick dynamic response. These output voltage signals are simulated based on the same output load current as the black-colored waveform.

V_{OUT} peak to peak value at the load transient with quick dynamic response is 288 mV and V_{OUT} peak to peak value at load transient without quick response is 354 mV. Thus, V_{OUT} peak to peak value with quick dynamic response is lower by 66 mV than without quick response. In addition, the settling time with quick dynamic response is shorter than without quick response.

Experimental results at the load transient with quick dynamic response (see Fig. 5.14) and without quick response (see Fig. 5.6) are as shown in Fig. 5.19. Without quick response cannot directly change its on-time width at the load transient. The blue-colored waveform represents the output voltage signal without quick response, and the red-colored waveform represents the output voltage signal with quick dynamic response. These output voltage signals are measured based on the same output load current as the black-colored waveform. The output load transient is generated by a function generator to generate fast slew rate control

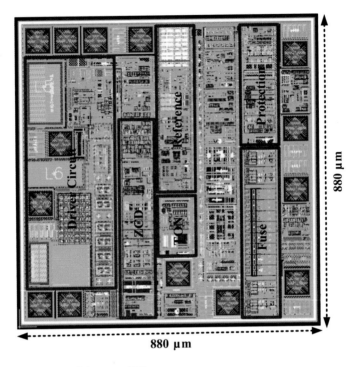

Fig. 5.16 Chip layout of the control IC

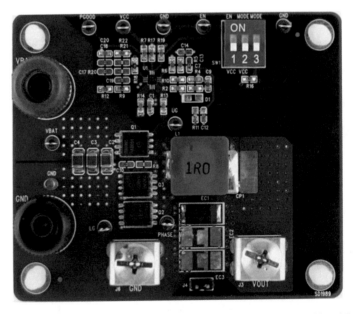

Fig. 5.17 Evaluation board of the constant frequency on-time control circuit with quick dynamic response for buck converter

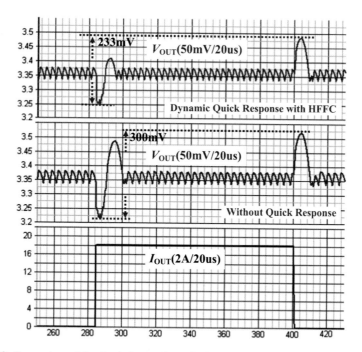

Fig. 5.18 Comparison of the simulation results at the load transient with quick dynamic response and without quick response

Fig. 5.19 Comparison of the experimental results at the load transient with quick dynamic response and without quick response

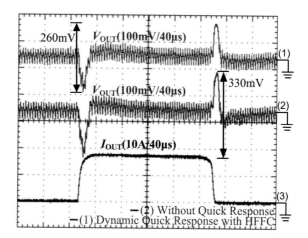

signal and drive the MOSFET switch. This implemented method produces faster load transient than the electronics load. V_{OUT} peak to peak value at the load transient with quick dynamic response is 260 mV, and V_{OUT} peak to peak value at the load transient without quick response is 340 mV. Thus, V_{OUT} peak to peak value variation with quick dynamic response is lower 80 mV than without quick

Fig. 5.20 Comparison of the experimental results at the droop with quick dynamic response and without quick response

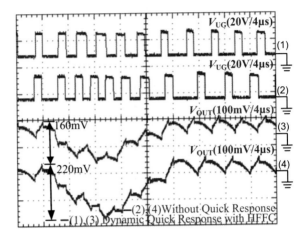

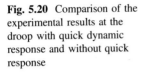

response. This quick dynamic response is useful to prevent the V_{OUT} from dropping significantly.

Experimental results at the droop with quick dynamic response and without quick response are as shown in Fig. 5.20. The blue-colored waveforms represent the output voltage and V_{UG} signal without quick response and the red-colored waveforms represent the output voltage and V_{UG} signal with quick dynamic response. The signals are measured based on the same output load current. V_{OUT} peak to peak value at the droop with quick dynamic response is 160 mV, and V_{OUT} peak to peak value at the droop without quick response is 220 mV. Thus, V_{OUT} peak to peak value with quick dynamic response is lower than without quick response, with a difference of about 60 mV. The setting time at the droop with quick dynamic response is also shorter than without quick response. In addition, the V_{UG} signal with quick dynamic response is obviously longer than without quick response at the droop, enabling the input power to deliver more energy to the output terminal.

Experimental results at the load release with quick dynamic response and without quick response are as shown in Fig. 5.21. The blue-colored waveforms represent the output voltage and V_{UG} signal without quick response and the red-colored waveforms represent the output voltage and V_{UG} signal with quick dynamic response. The signals are measured based on the same output load current. The overshoot of V_{OUT} peak to peak value at the load release with quick dynamic response is 80 mV and the overshoot of V_{OUT} peak to peak value at load release without quick response is 100 mV. In addition, the settling time at the load release with quick dynamic response is also shorter than without quick response because V_{OUT} does not suffer a voltage drop after load release.

The plans to base the identical operation conditions by comparing the external setting quick response for constant frequency on-time control circuit to understand the advantage and the superiority of this quick dynamic response further. Figure 5.22 compares the experimental results at the load transient with quick dynamic response (see Fig. 5.14) and external setting quick response (see

Fig. 5.21 Comparison of the experimental results at the load release with quick dynamic response and without quick response

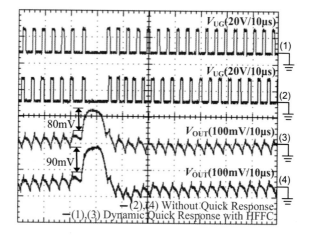

Fig. 5.22 Comparison of the experimental results at the load transient with quick dynamic response and external setting quick response

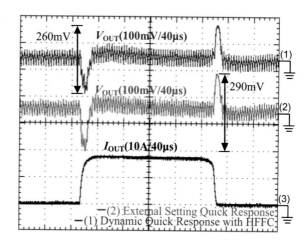

Fig. 5.12). The green-colored waveform represents the output voltage signal with external setting quick response, and the red-colored waveform represents the output voltage signal with quick dynamic response. These output voltage signals are measured based on the same output load current as the black-colored waveform. The output load current is used as a function generator to generate fast slew rate control signal, driving the MOSFET switch to implement the load transient. This implemented method is faster load transient than the electronics load.

V_{OUT} peak to peak value at the load transient with quick dynamic response is 260 mV and V_{OUT} peak to peak value at the load transient with external setting quick response is 290 mV. Thus, V_{OUT} peak to peak value with quick dynamic response and external setting quick response are close. Quick dynamic response and external setting quick response are useful to prevent the V_{OUT} from dropping significantly, but quick dynamic response does not require an extra pin to achieve the quick response function of IC.

Fig. 5.23 Comparison of the experimental results at the droop with quick dynamic response and external setting quick response

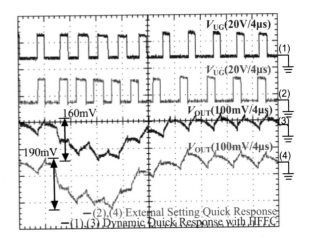

Experimental results at the droop with quick dynamic response and external setting quick response are as shown in Fig. 5.23. The green-colored waveforms represent the output voltage and V_{UG} signal with external setting quick response, and the red-colored waveforms represent the output voltage and V_{UG} signal with quick dynamic response. The signals are measured based on the same output load current.

V_{OUT} peak to peak value at the droop with quick dynamic response is 160 mV and V_{OUT} peak to peak value at the droop with external setting quick response is 190 mV. Thus, V_{OUT} peak to peak value with quick dynamic response and external setting quick response are also close, having a difference of only about 30 mV. In addition, the V_{UG} signal with quick dynamic response is obviously proportional to the output voltage drop. The V_{UG} signal with external setting quick response has the same QR-time width when the QR trigger circuit changes from a high-level signal to a low-level signal to turn OFF the S_2 switch.

5.5   Summary

This chapter proposes of the constant frequency on-time control circuit with quick dynamic response for Buck converter. The concept uses the high frequency feedback control to filters V_{OUT} at the load transient to change the on-time width dynamically. The input power can deliver more energy to the output terminal to prevent the V_{OUT} from dropping significantly.

Both experimental and simulation results confirm that the proposed with quick dynamic response can significantly improve the transient response of the constant frequency on-time control circuit. Moreover, the proposed quick dynamic response has a very simple structure and design of the constant frequency on-time control circuit for Buck converter.

References

1. C.J. Chen, D. Chen, C. W. Tseng, C. T. Tseng, Y. W. Chang, K.C. Wang, A novel ripple-based constant on-time control with virtual inductor current ripple for buck converter with ceramic output capacitors, in *Proceedings IEEE Applied Energy Conversion Congress and Exposition Conferences*, (2011), pp. 1244–1250
2. W.H. Ki, Signal flow graph in loop gain analysis of dc-dc PWM CCM switching converters. IEEE Trans. Circ. Syst. Part I. **6**, 644–654 (1998)
3. W.H. Ki, Analysis of subharmonic oscillation of fixed-frequency current-programming switch mode power converters. IEEE Trans. Circ. Syst. Part I. **45**(1), 104–108 (1998)
4. A.D. Schoenfeld, Y. Yu, ASDTIC control and standardized interface circuits applied to buck, parallel and buck-boost DC-to-DC power converters (NASA, Washington, DC, NASA Rep. NASA CR-121106, 1973)
5. C.W. Deisch, Switching control method changes power converter into a current source, in *Proceedings of IEEE Power Electronics Specialists Conferences*, (1978), pp. 300–306
6. P.L. Hunter, Converter circuit and method having fast responding current balance and limiting, U.S. Patent 4 002 963, 1 Nov 1977
7. L.H. Dixon, Average current-mode control of switching power supplies, in *Unitrode Power Supply Design Seminar Handbook*, (1990), pp. 5.1–5.14
8. N. Mohan, Power electronics circuits: an overview, in *Proceedings of IEEE Industrial Electronics Society Conference*, (1988), pp. 522–527
9. N. Mohan, W.P. Robbins, P. Imbertson, T.M. Undeland, R.C. Panaitescu, A.K. Jain, P. Jose, T. Begalke, Restructuring of first courses in power electronics and electric drives that integrates digital control. IEEE Trans. Power Electronics **18**, 429–437 (2003)
10. R.D. Middlebrook, S. Cuk, A general unified approach to modeling switching-converter power states, in *Proceedings of IEEE Power Electronics Specialists Conference*, (1976), pp. 73–86
11. D.Y. Chen, H.A. Owen, T. G. Wilson, Computer Aided design and graphics applied to the study of inductor-energy-storage dc-to-dc electronic power converters. IEEE Trans. Aerosp. Electron. Syst. **AES-9**, 585 (1973)
12. P. Burger, Analysis of a class of pulse modulated dc-to-dc power converters. IEEE Trans. Ind. Elect. Contr. Instrum. **IECI-22**, 104 (1975)
13. W.H. Lau, H. Chung, C.M. Wu, N.K. Poon, Realization of digital audio amplifier using zero-voltage-switched PWM power converter. IEEE Trans. Circ. Syst. Part I. **47**(3), 303–311 (2000)
14. C. Ni, T. Tetsuo, Adaptive constant on-time (D-CAP™) control study in notebook applications, *Texas Instruments, Application Report SLVA281B*, July, 2007
15. R. Redl, J. Sun, Ripple-based control of switching regulators an overview. IEEE Trans. Power Electron. **24**, 2669–2680 (2009)
16. J. Li, Current-mode control: modeling and its digital application. Ph.D. thesis, Virginia Polytechnic Institute and State University, 2009
17. J. Sun, Characterization and performance comparison of ripple-based control for voltage regulator modules. IEEE Trans. Power Electron. **21**, 346–353 (2006)
18. W. Huang, A new control for multi-Phase buck converter with fast transient response, in *Proceedings of IEEE Applied Power Electronics Conference and Exposition Conference*, (2001), pp. 273–279
19. J. Li, F.C. Lee, Modeling of V2 current-mode control, in *Proceedings of IEEE Applied Power Electronics Conference and Exposition Conferences*, (2009), pp. 298–304
20. S.J. Wang, Y.H. Lee, Y.C. Lai, K.H. Chen, Quadratic differential and integration technique in V2 control buck converter with small ESR capacitor, in *Proceedings of IEEE Custom Integrated Circuits Conference*, (2009), pp. 211–214
21. R. Redl, G. Reizik, Switched noise filter for the buck converter using the output ripple as the PWM ramp, in *Proceedings of IEEE Appl. Power Electron. Conference*, (2005), pp. 918–924

22. J. Li, F.C. Lee, New modeling approach and equivalent circuit representation for current-mode control. IEEE Trans. Power Electron. 1218–1230 (2010)
23. Y.H. Lee, S.J. Wang, K.H. Chen, Quadratic differential and integration technique in V2 control buck converter with small ESR capacitor. Proc. IEEE Trans. Power Electron. **25**, 829–838 (2010)
24. M.Y. Yen, P. Mok, A constant frequency output-ripple voltage-based buck converter without using large ESR capacitor. IEEE Trans. Circ. Syst. **55**, 748–752 (2008)
25. K.Y. Cheng, F. Yu, P. Mattavelli, F.C. Lee, Characterization and performance comparison of digital V^2-type constant on-time control for buck converters, IEEE Cont. Model. Power Electronics Conf. 1–6 (2010)
26. A.V. Petershevs, S.R. Sanders, Digital multimode buck converter control with loss-minimizing synchronous rectifier adaptation, IEEE Trans. Power Electron. **21**(6), 1588–1599 (2006)
27. S. Angkititrakul, H. Hu, Design and analysis of buck converter with pulse-skipping modulation, in *Proceedings of IEEE Power Electronics Specialists Conference*, (2008), pp. 1151–1156
28. X. Zhou, M. Donati, L. Amoroso, F.C. Lee, Improved light-load efficiency for synchronous rectifier voltage regulator module. IEEE Trans. Power Electronics **15**(5), 826–834 (2000)
29. C.L. Chen, W.L. Hsieh, W.J. Lai, K.H. Chen, C.S. Wang, A new PWM/PFM control technique for improving efficiency over wide load range, in *Proceedings of IEEE International Conference on Electronics, Circuits and Systems*, (Aug 2008), pp. 962–965
30. S. Kapat, S. Banerjee, A. Patra, Discontinuous map analysis of a DC-DC converter governed by pulse skipping modulation, IEEE Trans. Circ. Syst. Part I. **57**(7), 1793–1801 (2010)
31. W.W. Chen, J.F. Chen, T.J. Liang, L.C. Wei, J.R. Huang, W.Y. Ting, A novel quick response of RBCOT with VIC ripple for Buck converter. IEEE Trans. Power Electron. **28**, 4299–4308 (2013)
32. W.W. Chen, J.F. Chen, T.J. Liang, J.R. Huang, L.C. Wei, W.Y. Ting, Implementing dynamic quick response with high-frequency feedback control of the deformable constant on-time control for Buck converter on-chip. IET Power Electron. **6**(4), 383–391 (2013)
33. W.W. Chen, J.F. Chen, T.J. Liang, S.F. Hsiao, J.R. Huang, W.Y. Ting, Improved transient response using HFFC circuit of the CCRCOT with native AVP design for voltage regulators. IET Power Electron. **6**, 1948–1955 (2013)
34. W.W. Chen, J.F. Chen, T.J. Liang, J.R. Huang, W.Y. Ting, Improved transient response using HFFC in current-mode CFCOT control for buck converter, in *Proceedings IEEE International Conference on Power Electronics and Drive Systems (PEDS)*, (2013), pp. 546–549

Chapter 6
Ripple-Based Constant Frequency On-time Control Circuit with Virtual Inductor Current Ripple for Buck Converters

6.1 Challenges for Adaptive On-time Control Circuits for Buck Converters

Ceramic capacitors have many advantages for power converter applications. Ceramic capacitors are non-polarized. Electrolytic capacitors are polarized, resulting in damage or even explosion when subjected to high reversed polarity pulses or if they are mounted incorrectly. Ceramic capacitors have a wider frequency bandwidth and lower impedance than tantalums, making them more effective in suppressing noise in power lines. In the past, power converters needed two capacitors: a high-value electrolytic capacitor and a low-value ceramic capacitor. At present, high-value ceramic capacitors of the same size can be used. A single ceramic capacitor, which offers a better frequency response than the combination of electrolytic and ceramic capacitors, provides the required frequency response over a wide frequency band. Today's ceramic capacitors can do the entire task. In addition, ceramic capacitors have other advantages, particularly at high frequencies. Their ESR and ESL are much lower than those of tantalums. The ESR of ceramic capacitors is substantially lower than that of tantalum equivalents. A low ESR prevents overheating of the device and circuit, thus increasing the overall reliability.

Given these advantages, ceramic capacitors are highly suitable and preferred for many applications, such as digital cameras, netbooks, smartphones, and tablet computers due to their small size, low output voltage ripple, and high reliability requirements. Moreover, ceramic capacitors do not need temperature derating. In a circuit operating at 25 V and full-rated temperature, 25 V ceramic capacitors can be selected with full confidence.

However, ceramic capacitors contain a small output voltage ripple with a low ESR, which results in a small $R_{CO}*C_O$ time constant that makes it difficult to meet the critical design condition in Eq. (6.1). Thus, a small time constant results in an instability problem of subharmonic oscillations. The first challenge in adaptive on-time control circuits is the absence of virtual inductor current ripple to add to

© Springer Nature Singapore Pte Ltd. 2018
W.-W. Chen and J.-F. Chen, *Control Techniques for Power Converters with Integrated Circuit*, Power Systems,
https://doi.org/10.1007/978-981-10-7004-4_6

feedback voltage V_{FB}. This absence results in time-delay effects in the control loop. Thus, these control circuits still suffer from subharmonic oscillations with ceramic capacitors [1–13].

$$R_{CO} \cdot C_O > \frac{T_{ON}}{2} \tag{6.1}$$

The input signals of the comparator are reference voltage V_{REF} and feedback voltage V_{FB}. These signals can determine the output signal V_{TRIG} of the comparator. Reference voltage V_{REF} is generally designed to be a constant voltage less than 1 V, and it is without voltage ripple due to the regulation of the required and designed output voltage. If reference voltage V_{REF} has a voltage ripple, the output voltage would become large and cannot meet the original design. A low voltage ripple of feedback voltage V_{FB} can result in poor noise immunity between reference voltage V_{REF} and feedback voltage V_{FB}. Voltage signal V_{LX} easily suffers from the jitter phenomenon under poor noise immunity, which results in an unfixed switching frequency. A low-accuracy switching frequency causes the power converter or output device to encounter an error in operation. Hence, the second challenge in adaptive on-time control circuits is the low voltage ripple of feedback voltage V_{FB}, which results in the jitter phenomenon under poor noise immunity between reference voltage V_{REF} and feedback voltage V_{FB}. A virtual inductor current ripple can be added to feedback voltage V_{FB} to achieve improved noise immunity in the control loop. The virtual inductor current ripple exerts no impact on the output voltage ripple because the virtual inductor current ripple is directly added to feedback voltage V_{FB}.

Voltage signal V_{LX} is usually measured to determine its jitter performance, as shown in Fig. 6.1. Figure 6.1 shows voltage signal V_{LX} with the jitter phenomenon under poor noise immunity. To determine the jitter performance, one pulse of voltage signal V_{LX} can be triggered on the rising and falling edges by enabling a

Fig. 6.1 The voltage signal V_{LX} with a jitter phenomenon at the worse noise immunity

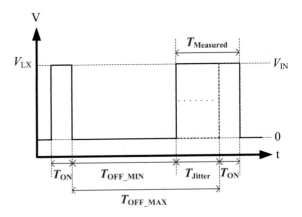

signal accumulation function of the oscilloscope. Then, the next pulse of the triggered pulse can contain a T_{Measured} time. If voltage signal V_{LX} has the jitter phenomenon, then T_{Measured} time has a longer width than T_{ON} time because T_{Measured} time is equal to the sum of T_{Jitter} time and T_{ON} time by Eqs. (6.2) and (6.3). The width of T_{Measured} time is equal to that of T_{ON} time when voltage signal V_{LX} does not exhibit the jitter phenomenon. With the jitter phenomenon, the maximum duty cycle D_{MAX} depends on $T_{\text{OFF\_MIN}}$, and the minimum duty cycle D_{MIN} depends on $T_{\text{OFF\_MAX}}$. D_{MAX} and D_{MIN} are calculated by Eqs. (6.4) and (6.5). Moreover, if the duty cycle considers the jitter phenomenon, the duty cycle can be changed and shown as Eq. (6.6). In general, the percentage of jitter ($\%_{\text{Jitter}}$) is used to estimate the jitter performance. $\%_{\text{Jitter}}$ depends on the D, D_{MAX}, and D_{MIN} by Eq. (6.7). The specification of $\%_{\text{Jitter}}$ is an uncommon definition in IC products' datasheet, but the internal specification of $\%_{\text{Jitter}}$ provided by IC Design Company is less than 15%.

$$T_{\text{Jitter}} = T_{\text{Measured}} - T_{\text{ON}} \tag{6.2}$$

$$T_{\text{OFF\_MIN}} = T_{\text{OFF\_MAX}} - T_{\text{Jitter}} \tag{6.3}$$

$$D_{\text{MAX}} = \frac{T_{\text{ON}}}{T_{\text{OFF\_MIN}} + T_{\text{ON}}} \tag{6.4}$$

$$D_{\text{MIN}} = \frac{T_{\text{ON}}}{T_{\text{OFF\_MAX}} + T_{\text{ON}}} \tag{6.5}$$

$$D = \frac{T_{\text{ON}}}{T_{\text{OFF\_MIN}} + T_{\text{ON}} + 0.5 * (T_{\text{Jitter}})} \tag{6.6}$$

$$\%_{\text{Jitter}} = \frac{D_{\text{MAX}} - D_{\text{MIN}}}{D} \tag{6.7}$$

Figure 6.2 shows the experimental results of voltage signal V_{LX} and voltage signal V_{OUT} with the jitter phenomenon under poor noise immunity for adaptive on-time control circuits. These experimental results were measured to trigger the rising edge of the first pulse by enabling a signal accumulation function of an oscilloscope. The next pulse of the triggered pulse can contain T_{Measured} time with the jitter phenomenon. T_{Measured} time is 664 ns, which includes many times of T_{ON} time. Pure T_{ON} time is the first pulse, and it is 200 ns. T_{Jitter} time is 464 ns. To determine the percentage of jitter ($\%_{\text{Jitter}}$), $T_{\text{OFF\_MIN}}$ time can be measured based on the experimental results. $T_{\text{OFF\_MIN}}$ time is 2216 ns, so $T_{\text{OFF\_MAX}}$ time can be calculated as 2680 ns. Based on Eqs. (6.3)–(6.5), D, D_{MAX}, and D_{MIN} can be calculated as 7.553, 8.278, and 6.944%, respectively. Then, $\%_{\text{Jitter}}$ can be obtained as 17.658%. This value is beyond the internal specification because it is larger than 15%.

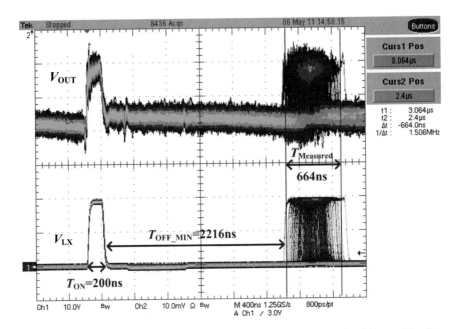

Fig. 6.2 Experimental results of the voltage signal V_{LX} and the voltage signal V_{OUT} with a jitter phenomenon at the worse noise immunity for the adaptive on-time control circuits

6.2   Implemented Control Circuits of the Victual Inductor Current Ripple

The victual inductor current ripple can be used to solve and improve the challenges in adaptive on-time control circuits, such as a small time constant that results in the instability problem with subharmonic oscillations and a low voltage ripple of feedback voltage V_{FB} that results in the jitter phenomenon under poor noise immunity between reference voltage V_{REF} and feedback voltage V_{FB}. Hence, the victual inductor current ripple must be added to feedback voltage V_{FB}.

Adding the inductor current ripple to feedback voltage V_{FB} is a good means to overcome loop stability problems because the inductor current ripple depends on the duty cycle, so it is a synchronous signal with feedback voltage V_{FB} ripple. In addition, the inductor current ripple is a sawtooth wave to be added to feedback voltage V_{FB} to increase the noise immunity and avoid time-delay effects in the control loop. For a large ESR such as that in electrolytic capacitors, feedback voltage V_{FB} ripple is similar to the inductor current ripple. For this reason, the control circuit also requires duty cycle information to generate the victual inductor current ripple.

In general, the control circuit uses two types of control signals to generate a victual inductor current ripple. One type of control signal is voltage signal V_{LX}, and

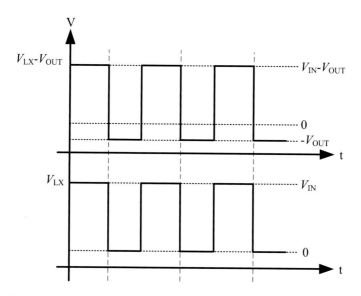

Fig. 6.3 Two types control signals to generate a victual inductor current ripple of the voltage signal V_{LX} and the voltage signal V_{LX}-V_{OUT}

the other type is the voltage drop between voltage signal V_{LX} and output voltage V_{OUT} (Fig. 6.3) because signal V_{LX} or the voltage drop between voltage signal V_{LX} and output voltage V_{OUT} have duty cycle information. These two types of control signals have the same peak-to-peak voltage, so they can be used to generate a victual inductor current ripple.

Moreover, the pulse control signal can use a low-pass filter to generate a sawtooth wave similar to the inductor current ripple. This sawtooth wave is called a victual inductor current ripple. The low-pass filter is placed between voltage signal V_{LX} and output voltage V_{OUT} or in voltage signal V_{LX} only. Two types of control circuits are typically used with the virtual inductor current ripple of adaptive on-time control circuits using a low-pass filter to connect voltage signal V_{LX} and output voltage V_{OUT}. One type is inductor's DCR current sensing control circuit [14–19]. Another one type of control circuit with virtual inductor current ripple of adaptive on-time control circuits using a low-pass filter to connect voltage signal V_{LX} is typically used; this type is ripple-based control circuit [1–5, 20–26].

The DCR current sensing control circuit is an alternative to a sense resistor, so a sense resistor does not need to be added. The DCR current sensing control circuit utilizes the parasitic resistance of an inductor to measure the current to the load. It remotely measures the current through an energy-storing inductor of a switching regulator circuit. Figure 6.4 shows the DCR current sensing control circuit of the adaptive on-time control circuits for buck converter. S_A and S_B are the switches, L is the output inductor, and R_{CO} is the ESR of the output capacitor C_O. The current source I_{OUT} is the output load current, and R_{D1} and R_{D2} are the feedback resistors to determine the output voltage. V_{FB} is a feedback voltage. The reference voltage V_{REF}

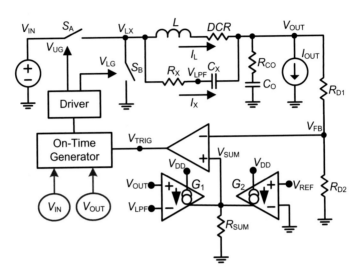

Fig. 6.4 The DCR current sensing control circuit of the adaptive on-time control circuits for buck converter

is created inside the IC. The on-time generator circuit samples the V_{IN} signal and V_{OUT} signal to adjust the on-time width to control the driver circuit and achieve the voltage regulation. A low-pass filter consists of a series of resistor R_X and capacitor C_X that are connected between voltage signal V_{LX} and output voltage V_{OUT}.

The most accurate sensing of the voltage across DCR is achieved by matching the time constant of resistor R_X and capacitor C_X filter with inductor L and its DCR time constant. The time constant of inductor L and its DCR are much larger than switching period T_S. The inductor current ripple contains a sawtooth wave. Switching period T_S is usually in the order of microseconds, given a switching frequency in the order of a few hundred kHz to a few MHz. The time constant of inductor L and its DCR are typically in the order of milliseconds, so the inductor current ripple always has a sawtooth wave. If the time constant of resistor R_X and capacitor C_X filter are selected to be equal to the time constant of inductor L and its DCR by Eq. (6.8), then voltage signal V_{CX} of capacitor C_X is directly proportional to the inductor current ripple, and voltage signal ripple V_{CX} is a virtual inductor current ripple. If the time constant of resistor R_X and capacitor C_X filter is not equal to the time constant of inductor L and its DCR, then the virtual inductor current ripple cannot obtain the same slew rate as the inductor current ripple.

$$R_X \cdot C_X = \frac{L}{DCR} \tag{6.8}$$

Voltage signal V_{CX} can be calculated by Eqs. (6.9)–(6.11). If a user wants to know DC voltage signal V_{CX}, it can be calculated by Eq. (6.12). Voltage signal V_{CX} is also equal to the drop between voltage signal V_{LPF} and output voltage V_{OUT}.

A real implemented IC uses two voltage-controlled current sources to add to the virtual inductor current ripple in feedback voltage V_{FB}, or the virtual inductor current ripple is subtracted in reference voltage V_{REF}, as shown in Fig. 6.4, because the same intent can be achieved. G_1 and G_2 are the gain of the voltage-controlled current source. They can directly control the amplitude of sum voltage ripple V_{SUM}.

$$I_L \cdot (s \cdot L + DCR) = I_X \cdot R_X + V_{CX} \tag{6.9}$$

$$I_X = s \cdot C_X \cdot V_{CX} \tag{6.10}$$

$$I_L \cdot (s \cdot L + DCR) = V_{CX} \cdot (1 + s \cdot R_X \cdot C_X) \tag{6.11}$$

$$V_{CX} = I_L \cdot DCR \tag{6.12}$$

Figure 6.5 shows these control signals with the DCR current sensing control circuit of adaptive on-time control circuits for buck converter. Reference voltage V_{REF} does not have a ripple-based circuit, so V_{REF} maintains a constant DC value.

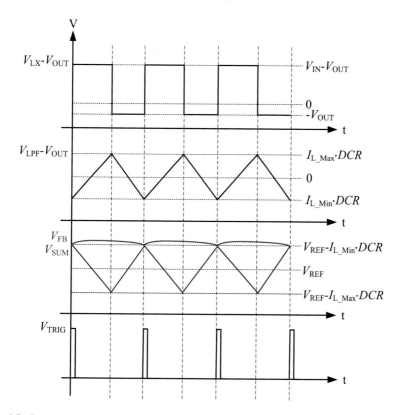

Fig. 6.5 Control signals with the DCR current sensing control circuit of the adaptive on-time control circuits for buck converter

The drop between voltage signal V_{LPF} and output voltage V_{OUT} can be controlled to become a sawtooth wave by the DCR current sensing control circuit. The DC level of V_A is equal to zero voltage, and V_A can be subtracted in reference voltage V_{REF} to increase the noise immunity and loop stability. If voltage V_{SUM} is larger than feedback voltage V_{FB} and control signal V_{TRIG} is changed from a low to high level, then control signal V_{TRIG} is needed to determine the final on-time width of T_{ON}. Output voltage V_{OUT} maintains the same voltage ripple even when the adaptive on-time control circuits use the DCR current sensing control circuit. The DCR current sensing control circuit is used not only to increase voltage ripple V_{REF} but also to maintain the same efficiency and output voltage ripple of the power converter.

However, the time constant of resistor R_X and capacitor C_X filter matching and DCR temperature dependence are important constraints that must be addressed to achieve a high level of current-sensing accuracy across variations in loads and operating temperatures. The DCR current-sensing control circuit can exhibit high performance, but the *DCR* parameter drifts when the temperature changes. Thus, *DCR* requires temperature compensation for current sensing in point-of-load regulator applications. In most IC products, the negative temperature coefficient (NTC) resistor is a good means to achieve temperature compensation. The NTC resistor should be added to the resistor network to monitor the inductor current for each phase and to achieve DCR thermal compensation [27–29]. Meanwhile, the DCR current-sensing control circuit needs to add two pins to achieve this function because the real implemented IC needs to sample voltage signal V_{LPF} and output voltage V_{OUT} individually. To achieve high accuracy in inductor current sensing, an auto tuning function must be implemented to overcome the impact of the inductor and its *DCR* tolerances. Thus, high-accuracy inductor current sensing with the DCR current-sensing control circuit is widely applied, such as in CPU voltage regulator applications.

The ripple-based control circuit is a type of control circuit with virtual inductor current ripple of adaptive on-time control circuits using a low-pass filter consisting of a series of resistor R_{LPF} and capacitor C_{LPF} that are connected to voltage signal V_{LX}. The ripple-based control circuit is also an alternative to a sense resistor, so adding a sense resistor is unnecessary. The ripple-based control circuit consists of R_{LPF}, C_{LPF}, and DC value extractor.

The ripple-based control circuit of the adaptive on-time control circuit with virtual inductor current ripple for buck converter is as shown in Fig. 6.6. S_A and S_B are the switches, L is the output inductor, and R_{CO} is the ESR of the output capacitor C_O. The current source I_{OUT} is the output load current, and R_{D1} and R_{D2} are the feedback resistors to determine the output voltage. V_{FB} is a feedback voltage. The reference voltage V_{REF} is created inside the IC. The on-time generator circuit samples the V_{IN} signal and V_{OUT} signal to adjust the on-time width to control the driver circuit and achieve the voltage regulation.

The purpose of a virtual inductor current ripple is to alleviate the instability problem because it enhances the effect of ESR voltage ripple in the feedback voltage. This circuit provides improved system stability, especially in all ceramic

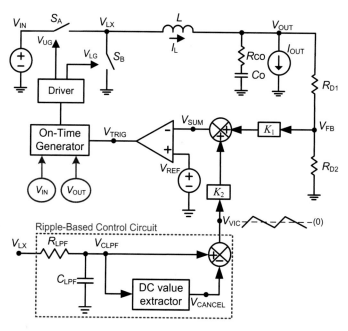

Fig. 6.6 The ripple-based control circuit of the adaptive on-time control circuit with virtual inductor current ripple for buck converter

capacitors for output capacitors, which normally have relatively low ESR values. In a ripple-based control circuit, voltage V_{CLPF} can be obtained by integrating voltage signal V_{LX} through R_{LPF} and C_{LPF} integrators. Voltage V_{CLPF} is a sawtooth wave, and its DC level is equal to the output voltage. Hence, V_{CLPF} cannot be directly added to feedback voltage V_{FB}. The DC value of V_{CLPF} is removed by the DC value extractor to generate the same voltage ripple with a zero DC value. One type of implemented circuit of the DC value extractor can be used with another low-pass filter to generate voltage V_{CANCEL}, which is similar to output voltage V_{OUT}. Voltage V_{CLPF} can subtract voltage V_{CANCEL} to obtain voltage V_{VIC} with the same ripple voltage as V_{CLPF}. However, its DC value is equal to zero voltage. Voltage V_{VIC} is added to feedback voltage V_{FB} to enhance the ESR voltage ripple.

The low-pass filter for R_{LPF} and C_{LPF} should be designed close to the resonant frequency of output inductor L and output capacitor C_O by Eq. (6.13). The equation needs to consider the minimum on-time T_{ON} by Eq. (6.14). Hence, the designs of resistor R_{LPF} and capacitor C_{LPF} need to meet the criteria for minimum on-time T_{ON} with a ripple-based control circuit. The design consideration of resistor R_{LPF} and capacitor C_{LPF} with a ripple-based control circuit is the similar to that of the D-CAP control circuit.

$$R_{\text{LPF}} \cdot C_{\text{LPF}} = \sqrt{L \cdot C_O} \qquad (6.13)$$

$$\frac{L \cdot C_O}{R_{\text{LPF}} \cdot C_{\text{LPF}}} > \frac{T_{\text{ON}}}{2} \qquad (6.14)$$

The main design consideration of the DC value extractor is to generate voltage V_{CANCEL}. This voltage obtains the same DC value as V_{OUT}. The time constant of the low-pass filter for the DC value extractor can affect its voltage ripple. If the time constant is too small, voltage V_{CANCEL} exhibits a large ripple. As a result, voltage V_{VIC} ripple cannot avoid the time-delay effects in the control loop. For this reason the time constant of the low-pass filter for the DC value extractor is approximately the time constant of resistor R_{LPF} and capacitor C_{LPF} filter to avoid the time-delay effects in the control loop. However, if the user considers increasing the transient response, the time constant of the low-pass filter for the DC value extractor can be designed to equal a half or less the time constant of resistor R_{LPF} and capacitor C_{LPF} filter because the virtual inductor current ripple is traded off to avoid time-delay effects in the control loop and to increase the transient response.

Figure 6.7 shows these control signals with the ripple-based control circuit of adaptive on-time control circuits for buck converters. The original feedback voltage $V_{\text{FB\_ORIGINAL}}$ does not have a ripple-based control circuit, so $V_{\text{FB\_ORIGINAL}}$ has a low voltage ripple, resulting in time-delay effects in the loop with a low ESR. Sum voltage V_{SUM} is the sum voltage of voltage V_{VIC} and original voltage $V_{\text{FB\_ORIGINAL}}$ with a ripple-based control circuit. V_{SUM} has a larger voltage ripple than the original voltage $V_{\text{FB\_ORIGINAL}}$ to increase noise immunity and loop stability because V_{SUM} ripple is similar to voltage V_{CLPF} ripple.

Output voltage V_{OUT} maintains the same waveform even when adaptive on-time control circuits employ a ripple-based control circuit. The ripple-based control circuit is used not only to increase the V_{SUM} ripple but also to maintain the same efficiency and output voltage ripple in the power converter.

A real implemented IC directly uses sum voltage V_{SUM} as the input terminal of the comparator, and the output voltage can be regulated and calculated by Eq. (6.15). Output voltage V_{OUT} is smaller than the required voltage, so the issue of the DC level of output voltage V_{OUT} being unequal to the required voltage needs to be solved.

$$V_{\text{OUT}} = (V_{\text{FB}} - 0.5 \cdot \Delta V_{\text{RIPPLE}}) \cdot \left(1 + \frac{R_{\text{D1}}}{R_{\text{D2}}}\right) \qquad (6.15)$$

If voltage V_{REF} is larger than sum voltage V_{SUM}, then control signal V_{TRIG} is changed from a low to high level. Control signal V_{TRIG} is needed to determine the final on-time width of T_{ON}. Moreover, a ripple-based control circuit can be embedded by a real implemented IC. Voltage signal V_{LX} also exists in IC, so no extra IC pin is needed to implement the modified control [1–13] because the ripple-based control circuit does not need to sample the output voltage signal. In

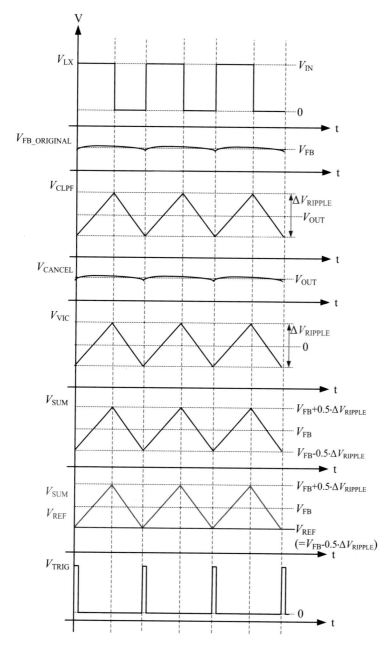

Fig. 6.7 Control signals with the ripple-based control circuit of the adaptive on-time control circuits for buck converter

addition, the real implemented IC should have a trimmed code option to change the suitable parameters of resistor R_{LPF}, capacitor C_{LPF}, and DC value extractor to meet different inductors and operation conditions.

6.3 On-time Generator Circuit of the Constant Frequency On-time Control Circuits for Buck Converters

On-time generator circuit of the constant frequency on-time control circuit is as shown in Fig. 6.8. The TON pin is in one of the IC, and the resistor R_1 can be placed between TON pin and input voltage to determine the width of on-time. V_{OUT} needs to be decided the peak voltage of capacitor C_1 by Eq. (6.16). The output voltage V_{OUT} divided by the input voltage V_{IN} is a duty cycle, and R_1, C_1, and G_1 are constant values, as shown in Eq. (6.17). G_1 is a gain of current-controlled current source. Even if the input voltage V_{IN} and output voltage V_{OUT} are changed, the system still maintains the same switching frequency [1–5].

$$T_{ON} = \frac{V_{OUT}}{V_{IN}} \times \frac{R_1 \times C_1}{G_1} \tag{6.16}$$

$$T_{ON} = \frac{V_{OUT}}{V_{IN}} \times T_S \tag{6.17}$$

Constant frequency on-time control circuit is a different from the conventional COT control circuit because it avoids the generation of the same on-time width at a high V_{OUT}. Regardless of changes in V_{IN} or V_{OUT}, the conventional COT control circuit still generates the same on-time width. If the conventional COT control circuit wants to regulate a high V_{OUT} and needs to generate more on-time pulses, an increase in switching loss occurs. Thus, the constant frequency on-time control circuit is suitable for a wide range output voltage.

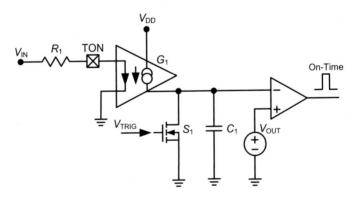

Fig. 6.8 On-time generator circuit of the constant frequency on-time control circuit

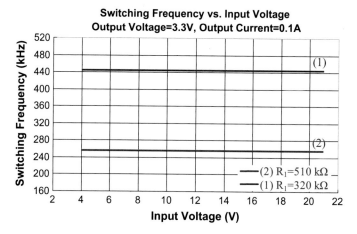

Fig. 6.9 Experimental results of the switching frequency versus the input voltage for constant frequency on-time control circuit

Experimental results of the switching frequency versus the input voltage for constant frequency on-time control circuit are as shown in Fig. 6.9. The operation conditions are set at 3.3 V output voltage and 0.1 A output current to measure the switching frequency for 4–21 V input voltage. When R_1 is equal to 510 kΩ, connecting V_{IN} and TON pin, the switching frequency is 255 kHz, regardless of changes in V_{IN}. If R_1 is reduced to 320 kΩ, the switching frequency changes from 255 to 444 kHz and the resistor R_1 directly affects the on-time width. Between the resistor R_1 and the switching frequency are the inverse relationships.

Experimental results of the inductor current ripple versus the input voltage for constant frequency on-time control circuit are as shown in Fig. 6.10. The operating conditions are set 3.3 V output voltage and 0.1 A output current to measure the inductor current ripple for 4–21 V input voltage. If the inductance is invariant, then the inductor current ripple is related to the voltage drop between the input voltage and output voltage and is also dependent on the on-time T_{ON}, as shown by Eq. (6.18). By substituting Eq. (6.17) into Eq. (6.18), the inductor current ripple is obtained as Eq. (6.19). The inductor current ripple increases as the input voltage is increased. Hence, it does not maintain a constant value. When R_1 is increased from 320 to 510 kΩ, the on-time width is also increased, resulting in a larger inductor current ripple.

$$\Delta I_L = \frac{V_{IN} - V_{OUT}}{L} \times T_{ON} \tag{6.18}$$

$$\Delta I_L = \frac{V_{OUT}(1 - (V_{OUT}/V_{IN}))}{L} \times T_S \tag{6.19}$$

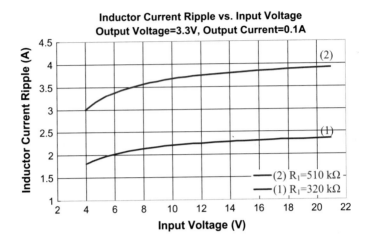

Fig. 6.10 Experimental results of the inductor current ripple versus the input voltage for constant frequency on-time control circuit

6.4 Comparison of Quick Dynamic Response and Conventional Quick Response of the On-time Generator Circuit

Two types of improved transient response circuits are typically used, namely, internal fixed by IC design and external setting by component or voltage source. The internal fixed by IC design usually uses the feedback voltage V_{FB} to trigger the quick response function, which can provide a long on-time width to control the driver circuit. However, the generator circuit of quick response has a very similar on-time generator circuit. Figure 6.11 shows the internal fixed quick response of the on-time generator circuit for constant frequency on-time control circuit [1–5].

The internal fixed quick response needs to add a voltage source V_{REF2}. The V_{REF2} is usually designed at 80–90% of V_{REF} to prevent the occurrence of under voltage protection. If the feedback voltage V_{FB} is lower than V_{REF2}, the output signal of the comparator changes from high-level to low-level, which can turn off the switch of S_2. The voltage source V_{REF1} is set in series connection with V_{OUT} to increase the on-time width. V_{REF1} is designed by the maximum duty cycle limited [1–5].

The main advantage of this implementation is that no extra pin in the IC is needed to achieve the quick response function. However, the design of the internal fixed quick response can generate just one form of a long on-time width. Thus, this circuit cannot dynamically change the on-time width during different load conditions. Besides, the hysteresis of the comparator is difficult to design, especially for a very fast load transient. If the hysteresis of the comparator is too small, it can induce the system to erroneously trigger a long on-time width, thereby causing high output

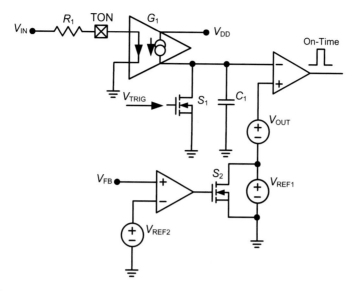

Fig. 6.11 Internal fixed quick response of the on-time generator circuit for constant frequency on-time control circuit

voltage. If the hysteresis of the comparator is too large, it can also render the quick response function unavailable.

External setting quick response of the on-time generator circuit for constant frequency on-time control circuit is as shown in Fig. 6.12. The operation principle of the external setting quick response requires the addition of a voltage source V_{QRSET} to determine the QR-time width. V_{QRSET} is connected to the QRSET pin. If the QR-time needs a long QR-time width, V_{QRSET} should be larger than V_{OUT}.

V'_{OUT} determines the peak voltage of C_1. Thus, if V'_{OUT} is high, the peak voltage increases, causing a long on-time width. V'_{OUT} is implemented to provide a ripple-less voltage because it can keep the switching frequency constant without under output voltage ripple and noise interference. V'_{OUT} is implemented to sample the Phase signal through a second-order low-pass filter, producing a similar V_{OUT} signal. Therefore, V'_{OUT} is ripple-less and has slower transient response than the V_{OUT} signal.

The QR trigger circuit samples V_{OUT} and uses the low-pass filter to cause V_{OUT} delay. This trigger circuit also needs the voltage signal V_{QRTH}. In the steady-state operation, the V_{OUT} drop cannot trigger the QR-time. When this abrupt voltage drop is lower than V_{QRTH}, the QR trigger circuit changes from a high-level signal to a low-level signal to turn OFF the S_2 switch and then the on-time generator circuit generates the QR-time width. The low-pass filter frequency for R_{LOW} and C_{LOW} should be designed much smaller than the switching frequency which can avoid this system to fail in its operation with QR-time. V_{QRTH} can be designed in the IC or set by the user through the QRTH pin. The main advantage of the external setting

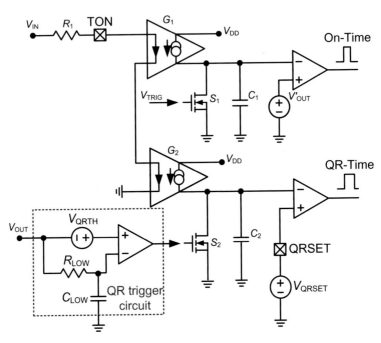

Fig. 6.12 External setting quick response of the on-time generator circuit for constant frequency on-time control circuit

quick response is that it can depend on the load conditions to design the V_{QRSET} value, generating a suitable QR-time width in this system. However, this method requires two extra IC pins to achieve the quick response function [1–5].

Quick dynamic response of the ripple-based constant frequency on-time control circuit with virtual inductor current ripple for buck converter is proposed, as well as quick dynamic response of on-time generator circuit for constant frequency on-time control circuit, as shown in Fig. 6.13. An external setting device is used to design these components, instead of a set-up voltage source. This quick dynamic response does not require an extra pin to achieve the quick response function of IC. The system consists of a series of the resistor R_q and capacitor C_q that are connected between the output voltage V_{OUT} and TON pin. It operates under the principle that the high-pass filter filters high frequency signal to pass from V_{OUT} to the TON pin by Eq. (6.20). However, the frequency of the high-pass filter must be larger than or equal the switching frequency so the steady state operation of the system is not affected. On the other hand, the user may base the design of the resistor R_q and capacitor C_q on the worst-case operation (maximum loading step) [1–5].

V'_{OUT} determines the peak voltage of C_1. Thus, if V'_{OUT} is high, the peak voltage increases, causing a long on-time width. V'_{OUT} is implemented to provide a ripple-less voltage because it can keep the switching frequency constant without under output voltage ripple and noise interference. V'_{OUT} is implemented to sample

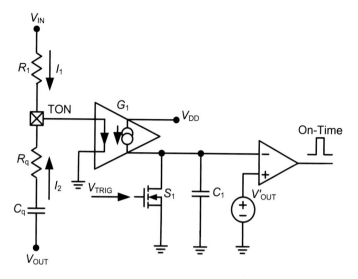

Fig. 6.13 Quick dynamic response of the on-time generator circuit for constant frequency on-time control circuit

the Phase signal through a second-order low-pass filter producing a similar V_{OUT} signal. Therefore, V'_{OUT} is ripple-less and has a slower transient response than V_{OUT}.

$$F_{RC} = \frac{1}{2\pi \times R_q \times C_q} \geq F_S \tag{6.20}$$

When light load quickly transforms to heavy load, V_{OUT} momentarily drops through the coupling by the capacitor C_q, which induces the resistor R_q to cause a voltage drop. The voltage drop of R_q will form a current of I_2. Thus, the resistor R_q needs to be designed first because the current of I_2 can directly change the width of on-time. However, a longer width of on-time induces the converter to deliver more energy from the input terminal to the output loading. If the resistor R_q has reduced, the width of on-time becomes longer. The capacitor of C_q should be designed with the frequency of the high-pass filter. The calculation of the Laplace transform formula is shown in Eq. (6.21).

$$T_{ON}(s) = \frac{C_1}{G_1\left((V_{IN}/R_1) + (sC_q V_{OUT})/(1 + sC_q R_q)\right)} \times V'_{OUT} \tag{6.21}$$

The advantages of this quick dynamic response are as follows: (1) It clearly generates a longer width of on-time that is proportional to the output voltage drop; (2) Adaption to the width of on-time depends on the load conditions of the system; (3) It only uses one on-time generator circuit; thus, it is very convenient to design and apply; (4) It does not require an extra pin to achieve the quick response circuit function of IC.

6.5 Experimental Results

The experimental results are shown to prove the feasibility and performance with quick dynamic response of the ripple-based constant frequency on-time control circuit with virtual inductor current ripple for buck converter. The specifications are as follows:

(1) Input voltage (V_{IN}): 12 V
(2) Output voltage (V_{OUT}): 3.3 V
(3) Maximum output load current (I_{OUT}): 18 A @ 1.2857 A/μs
(4) Switching frequency (F_S): 255 kHz
(5) MOSFET (S_A, S_B): BSC0909NS * 3
(6) Feedback resistors (R_{D1}, R_{D2}): 68 and 12 kΩ
(7) Main inductor (L): IHLP4040DZER1R0MA1 (1 μH)
(8) Output capacitors (C_O): 22μF/6.3 V (R_{CO}: 3 mΩ)*13
(9) Reference voltage (V_{REF}): 0.5 V
(10) Quick dynamic response: $G_1 = 1(A/A)$; $R_1 = 510$ kΩ; $C_1 = 7.06$ pF; $R_q = 510$ Ω; $C_q = 390$ pF;

Figure 6.14 shows the chip layout of the control IC, in which the low-pass filter of the virtual inductor current ripple occupies the area marked by the LP Filter with a die size of 780 μm × 780 μm.

Experimental results at the load transient with quick dynamic response (see Fig. 6.13) and without quick response (see Fig. 6.8) are as shown in Fig. 6.15. The blue-colored waveform represents without quick response for the output voltage

Fig. 6.14 Chip layout of the control IC

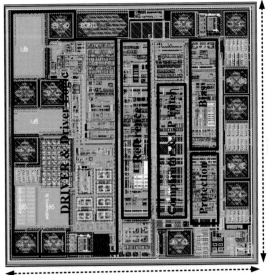

780 μm

780 μm

Fig. 6.15 Comparison of the experimental results at the load transient with quick dynamic response and without quick response

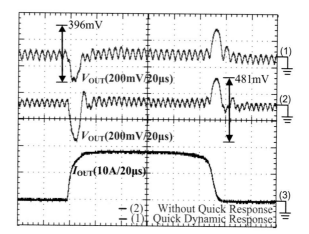

signal, whereas the red-colored waveform represents with quick dynamic response for the output voltage signal. V_{OUT} peak to peak value at the load transient with quick dynamic response is 396 mV, and V_{OUT} peak to peak value at the load transient without quick response circuit is 481 mV. Thus, V_{OUT} peak to peak value with quick dynamic response is lower by 85 mV than without quick response, which is useful to prevent the V_{OUT} from dropping significantly.

The output voltage signals are measured based on the same output load current, such as the black-colored waveform. The output load current is used as a function generator signal to control the switch of MOSFET to implement the load transient, which is faster than the electronics load as shown in Fig. 6.16.

Experimental results at the droop with quick dynamic response and without quick response are as shown in Fig. 6.17. The blue-colored waveforms represent without quick response circuit for the output voltage and V_{UG} signal, whereas the red-colored waveforms represent with quick dynamic response for the output

Fig. 6.16 Comparison of the slew rate at the load transient for the electronics load and MOSFET implemented with a function generator signal

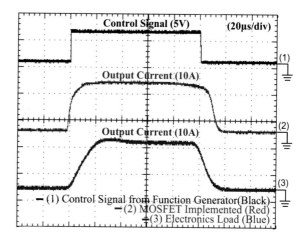

Fig. 6.17 Comparison of the experimental results at the droop with quick response dynamic and without quick response

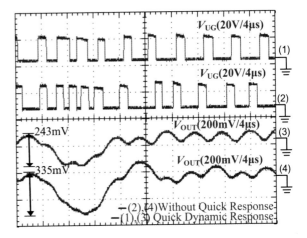

voltage and V_{UG} signal, whose signals are measured based on the same output load current. V_{OUT} peak to peak value at the droop with quick dynamic response is 243 mV, and V_{OUT} peak to peak value at the droop without quick response circuit is 335 mV. Thus, V_{OUT} peak to peak value with quick dynamic response is lower by 92 mV than without quick response, which is useful to prevent the V_{OUT} from dropping significantly. The setting time with quick dynamic response at the droop is faster than without quick response, because V_{OUT} at the droop with quick dynamic response does not suffer an overshoot. On the other hand, the V_{UG} signal with quick dynamic response is obviously longer than without quick response circuit at the droop, which enables the input power to deliver more energy to the output terminal.

Experimental results at the load release with quick dynamic response and without quick response are as shown in Fig. 6.18. The blue-colored waveforms represent without quick response for the output voltage and V_{UG} signal, whereas the red-colored waveforms represent with quick dynamic response for the output voltage and V_{UG} signal, whose signals are measured based on the same output load current. V_{OUT} at the load release with quick dynamic response varies closely without quick response. The settling time at the load release with quick dynamic response is also shorter than without quick response because V_{OUT} does not suffer a voltage drop.

To further understand the advantage and superiority of this quick dynamic response, the plans based the same V_{OUT} peak to peak value at the droop with quick dynamic response on to increase additional 22 µF of ceramic output capacitors without quick response.

The conditions are as follows:

Condition 1: Quick dynamic response

(1) Output capacitors C_O: 22 µF/25 V (R_{CO}: 3 mΩ) × 13
(2) Quick dynamic response: $G_1 = 1$(A/A); $R_1 = 510$ kΩ; $C_1 = 7.06$ pF; $R_q = 510$ Ω; $C_q = 390$ pF;

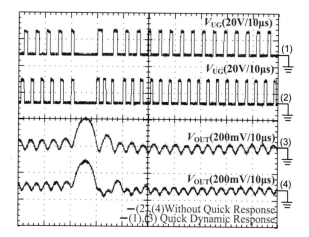

Fig. 6.18 Comparison of the experimental results at the load release with quick dynamic response and without quick response

Condition 2: Without quick response using extra 17×22 µF ceramic capacitors

(1) Output capacitors C_O: 22 µF/25 V (R_{CO}: 3 mΩ) $\times$ 30

Based on the same V_{OUT} peak to peak value at the droop with quick dynamic response, it needs to add how many 22 µF of output ceramic capacitors without quick response. Figure 6.19 shows a comparison of the experimental results at the load transient with quick dynamic response and without quick response using extra 17×22 µF of ceramic output capacitors. The blue-colored waveform represents without quick response with 30×22 µF ceramic output capacitors for the output voltage signal, whereas the red-colored waveform represents with quick dynamic response with 13×22 µF ceramic output capacitors for the output voltage signal, whose signals are measured based on the same output load current. Thus, the large output capacitance can reduce the V_{OUT} ripple at the load transient, but the cost and size of the output capacitance need to be sacrificed. On the other hand, the large

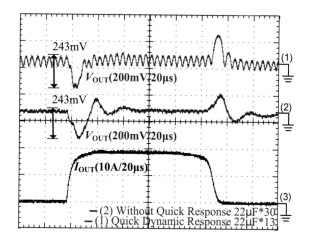

Fig. 6.19 Comparison of the experimental results at the load transient with quick dynamic response and without quick response using extra 17×22 µF of ceramic output capacitors

output capacitance induces the system loop response slowly. The setting time at load transient with quick dynamic response is faster than without quick response because V_{OUT} at the droop with quick dynamic response does not suffer an overshoot and V_{OUT} at the load release with quick dynamic response does not suffer a voltage drop.

Based on the same V_{OUT} peak to peak value at the droop with quick dynamic response, it needs to set additional 17×22 μF of ceramic output capacitors without quick response. Thus, quick dynamic response at this operating condition, which can save 17×22 μF ceramic output capacitors, regardless of the cost and size of the output capacitance, both of which are the very benefits of the circuit design.

Experimental results at the droop with quick dynamic response and without quick response using extra 17×22 μF of ceramic output capacitors are as shown in Fig. 6.20. The blue-colored waveforms represent without quick response with 30×22 μF ceramic output capacitors for the output voltage and V_{UG} signal, whereas the red-colored waveforms represent with quick dynamic response with 13×22 μF ceramic output capacitors for the output voltage and V_{UG} signal, whose signals are measured based on the same output load current. The setting time with quick dynamic response at the droop is 38 μs, which is faster by 25.5 μs than without quick dynamic response, because V_{OUT} at the droop with quick dynamic response does not suffer an overshoot. Thus, the large output capacitance can hold the V_{OUT} drop and induce the V_{OUT} ripple small. However, it may cause the system to obtain a slow transient response.

Figure 6.21 shows a comparison of the experimental results at the load release for the quick dynamic response and without quick response. The blue-colored waveforms represent without quick response with 30×22 μF ceramic output capacitors for the output voltage and the V_{UG} signal, whereas the red-colored waveforms represent the quick dynamic response with 13×22 μF ceramic output capacitors for the output voltage and the V_{UG} signal, whose signals are measured based on the same output load current.

Fig. 6.20 Comparison of the experimental results at the droop with quick dynamic response and without quick response using extra 17×22 μf of ceramic output capacitors

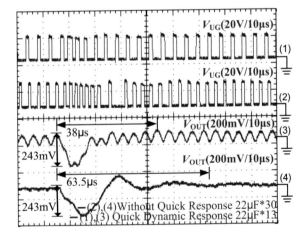

Fig. 6.21 Comparison of the experimental results at the load release with quick dynamic response and without quick response using extra 17 × 22 μF of ceramic output capacitors

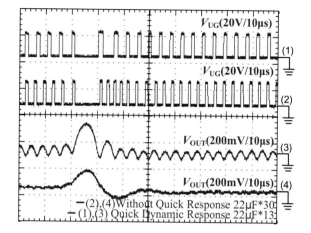

Thus, the large output capacitors can reduce the peak voltage of V_{OUT} at the load release, but the cost and size of the output capacitors need to be sacrificed. On the other hand, the large output capacitors induce the system loop response slowly. The setting time at load release with the quick dynamic response is faster than without quick response.

6.6 Summary

This chapter proposes quick dynamic response of the ripple-based constant frequency on-time control circuit with virtual inductor current ripple for buck converter. Using the capacitor and resistor in series to filter V_{OUT} at load transient to dynamically change the width of on-time, the input power delivers more energy to the output terminal to prevent V_{OUT} from dropping significantly.

Experimental results confirm quick dynamic response which can reduce in cost and size of the ceramic output capacitance and improve in transient response. Moreover, the proposed quick dynamic response has a very simple structure and design of the ripple-based constant frequency on-time control circuit with virtual inductor current ripple for buck converter.

References

1. W.W. Chen, J.F. Chen, T.J. Liang, L.C. Wei, J.R. Huang, W.Y. Ting, A novel quick response of RBCOT with VIC ripple for buck converter. IEEE Trans. Power Electron. **28**, 4299–4308 (2013)
2. W.W. Chen, J.F. Chen, T.J. Liang, J.R. Huang, L.C. Wei, W.Y. Ting, Implementing dynamic quick response with high-frequency feedback control of the deformable constant on-time control for Buck converter on-chip. IET Power Electron. **6**(4), 383–391 (2013)

3. W.W. Chen, J.F. Chen, T.J. Liang, S.F. Hsiao, J.R. Huang, W.Y. Ting, Improved transient response using HFFC circuit of the CCRCOT with native AVP design for voltage regulators. IET Power Electron. **6**, 1948–1955 (2013)
4. W.W. Chen, J.F. Chen, T.J. Liang, J.R. Huang, W.Y. Ting, Improved Transient Response Using HFFC in Current-Mode CFCOT Control for Buck Converter, in *Proceedings of IEEE International Conference on Power Electronics and Drive Systems (PEDS)*, (2013), pp. 546–549
5. C.J. Chen, D. Chen, C.W. Tseng, C.T. Tseng, Y.W. Chang, K.C. Wang, A novel ripple-Based constant On-Time control with virtual inductor current ripple for buck converter with ceramic output capacitors, in *Proceedings of IEEE Applied Energy Conversion Congress and Exposition conference*, (2011), pp. 1244–1250
6. R. Redl, J. Sun, Ripple-based control of switching regulators an overview. IEEE Trans. Power Electron. **24**, 2669–2680 (2009)
7. J. Sun, Characterization and performance comparison of ripple-based control for voltage regulator modules. IEEE Trans. Power Electron. **21**, 346–353 (2006)
8. W. Huang, A new control for multi-phase buck converter with fast transient response, in *Proceedings of IEEE Applied Power Electronics Conference and Exposition conference*, (2001), pp. 273–279
9. J. Li, F.C. Lee, Modeling of V2 current-mode control, in *Proceedings of IEEE Applied Power Electronics Conference and Exposition conference*, (2009), pp. 298–304
10. K.D.T. Ngo, S.K. Mishra, M. Walters, Synthetic-ripple modulator for synchronous buck converter. Proceed. IEEE Power Electron. **3**, 148–151 (2005)
11. Y.H. Lee, S.J. Wang, K.H. Chen, Quadratic differential and integration technique in V2 control buck converter with small ESR capacitor. Proc. IEEE Trans. Power Electron. **25**, 829–838 (2010)
12. M.Y. Yen, P. Mok, A constant frequency output-ripple voltage-based buck converter without using large ESR capacitor. IEEE Transactions on Circuits and Systems **55**, 748–752 (2008)
13. K.Y. Cheng, F.Yu, P. Mattavelli, F.C. Lee, Characterization and performance comparison of digital V^2-type constant on-time control for buck converters, in *IEEE Control and Modeling for Power Electronics conference*, (2010), pp. 1–6
14. Richtek Tech. Corp., Comparison of DCR current sense topologies, Application Note, 2015
15. Texas Instruments Inc., TPS65311-Q1 BUCK1 controller DCR current sensing, Application Report, 2016
16. S. Saggini, A. Zafarana, D. Zambotti, M. Ghioni, Digital autotuning system for inductor current sensing in voltage regulation module applications. IEEE Trans. Power Electron. **23**(5) (2008)
17. Y. Yan, F.C. Lee, P. Mattavelli, Unified three-terminal switch model for current mode controls. IEEE Trans. Power Electron. **27**(9), 4060–4070 (2012)
18. L. Hua, S. Luo, Design considerations for small signal modeling of DC-DC converters using inductor dcr current sensing under time constants mismatch conditions, in *Power Electronics Specialists Conference, 2007, PESC'07*, (2007), pp. 2182–2188
19. Y. Sun, J. Li, F.C. Lee, Modeling of digitally controlled voltage regulator modules, in *Applied Power Electronics Conference and Exposition, 2010, APEC '10*, (2010), pp. 176–182
20. Texas Instruments Inc., "D-CAP™ Mode with all-ceramic output capacitor application, Application Report, 2011
21. S. J. Wang, Y.H. Lee, Y.C. Lai, K.H. Chen, Quadratic differential and integration technique in V2 control buck converter with small ESR capacitor, in *Proceedings IEEE Custom Integrated Circuits Conference*, (2009), pp. 211–214
22. R. Redl, G. Reizik, Switched noise filter for the buck converter using the output ripple as the PWM ramp, in *Proceedings IEEE Applied Power Electronics Conference*, (2005), pp. 918–924
23. R. Redl, T. Schiff, *A new family of enhanced ripple regulators for power-management applications* (Nuremburg, Germany, International Exhibition and Conf. Eur., 2008), pp. 255–268

24. Richtek Tech. Corp., "3A, 18 V, 700 kHz ACOT^{TM} synchronous step-down converter, RT7275/RT7276 Datasheet, 2017
25. Texas Instruments Inc., TPS54325 4.5-V to 18-V, 3-A Output synchronous step down switcher with integrated FET, TPS54325 Datasheet, 2014
26. Texas Instruments Inc., Single-phase, D-CAP™ and D-CAP2™ Controller with 2-Bit flexible VID control, TPS51518 Datasheet, 2011
27. Richtek Tech. Corp., DCR Temperature Compensation," Application Note, 2014
28. Richtek Tech. Corp., Multi-Phase PWM controller for CPU core power supply, RT8859 M Datasheet, 2014
29. Richtek Tech. Corp., Dual output 3-Phase + 2-Phase PWM controller for CPU and GPU Core power supply, RT8885A Datasheet, 2012

Chapter 7
Constant Current Ripple On-Time Control Circuit With Native Adaptive Voltage Positioning Design for Voltage Regulators

7.1 Challenges for Voltage Regulators

In recent years, over one billion transistors have been integrated in one processor, core static current has been increased from 20 to 100 A, and core voltage has been reduced from 2 to 0.7 V [1–3]. It is a challenge to provide the large output loading requirement for CPU application [4–8].

Single-phase VR can be widely used in low-voltage converter applications with an output loading of up to approximately 25 A. However, power dissipation, power stress of the components, and efficiency become an issue under a large output load current. A suitable approach is to use multiphase VRs, as shown in Fig. 7.1. The benefits of using multiphase VRs versus single-phase VRs and the value of multiphase VRs become evident when they are implemented.

In multiphase VRs, the phases are shifted based on the number of power trains. For example, a three-phase converter is phase shifted by 120° (0°, 120°, and 240°) and so on. Moreover, each phase needs to have a minimum individual off-time limit. The interleaved control also reduces the output voltage ripple effectively. Hence, the phases are interleaved. Interleaving effectively reduces ripple currents at the input and output terminals. It also reduces hot spots on a PCB or a component. In effect, two-phase VRs reduce the RMS current power dissipation in the power MOSFET and inductors by half. Interleaving also reduces transitional losses.

The output filter requirements decrease in the implementation of multiphase VRs due to the reduced current in the power stage for each phase. For an operation involving 100 A of output load current, an average current of only 33 A is delivered to each inductor when multiphase VRs are adopted. Compared with using a single-phase approach for this 100 A scenario, the inductance and inductor size are drastically reduced because of low average and saturation currents. The ripple current of output load cancellation in the output-filter stage in multiphase VRs results in a reduced ripple voltage across the output capacitor compared with the

© Springer Nature Singapore Pte Ltd. 2018
W.-W. Chen and J.-F. Chen, *Control Techniques for Power Converters with Integrated Circuit*, Power Systems,
https://doi.org/10.1007/978-981-10-7004-4_7

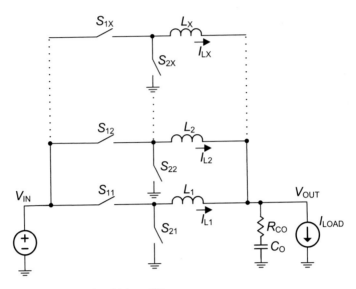

Fig. 7.1 Control structure of multiphase VRs

situation in a single-phase VR. For these reasons, multiphase converters are preferred over single-phase ones.

Recent CPU developments have enhanced power loss reduction when the CPU
is on standby. To improve power loss, it is must reduce switching loss and conduction loss when the load is light. The control scheme of the adaptive on-time of
VRs can reduce switching loss and then increase light load efficiency [9–23]
because the switching frequency depends on the loading condition. If the load is
light, the switching frequency is reduced to increase efficiency. On the other hand,
the PSM mode [24–29] can be combined with the constant on-time to increase light
load efficiency because the PSM function turns off the lower-side switch when the
inductor current is equal to 0 A. It not only reduces switching loss, but also saves
system conduction loss.

The other challenge is to reduce the output capacitors, the bill of material size,
and the bill of material cost. Most of VRs have an AVP function to improve or to
solve this challenge [3, 9, 30–37]. AVP concept has been widely used in VRs of the
CPU application for energy saving in CPU power. As shown in Fig. 7.2, the
steady-state target load-line with ideal AVP for VRs, the steady-state output voltage
is lower at full-load. This reduces the power loss and heat problem of CPU in
full-load.

Output voltage waveforms when the load current goes through step changes with
ideal AVP for VRs are as shown in Fig. 7.3. The load-line is bounded by the
tolerance band indicated by the two dotted lines which is about 38 mV from Intel's
VR12.0 specification. When the load changes, it is inevitable that output voltage
jumps but the voltage must stay within the tolerance band. If output voltage goes

Fig. 7.2 Steady-state target load-line with ideal AVP for VRs

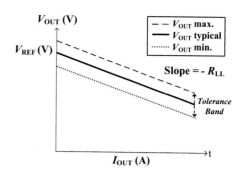

Fig. 7.3 Output voltage waveforms for step-load changes with ideal AVP for VRs

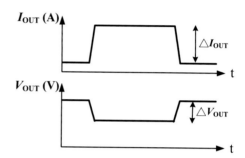

above upper tolerance band, CPU may have heat problem; if output voltage goes below lower tolerance band, computer may shut down.

Another benefit of AVP concept is the reduction of the output voltage deviation during load transient. Figure 7.4 shows the load transient response of CPU VRs with and without AVP. For a conventional control without AVP, output voltage is regulated at a fixed level. On the other hand, when AVP is implemented, output voltage stays in the voltage near upper tolerance band after load release. As can be seen from Fig. 7.4, the voltage deviation for control with AVP is about half of the control without AVP. This relaxes the required number of output capacitors in parallel to reduce equivalent ESR value while increase equivalent capacitance. The reduction of capacitor number is meaningful when considering the large percentage

Fig. 7.4 Load transient response of CPU VRs with and without AVP

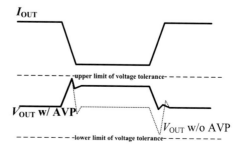

Fig. 7.5 Equivalent circuit of a VR with ideal AVP

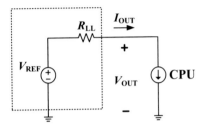

of area occupied by VRs components in computer motherboard. Besides, implementation of this concept not only reduces the power converter output capacitor requirement but also lowers the power consumption in the microprocessor loads at full load.

It is pointed out that to achieve ideal AVP. The constant small-signal output impedance versus frequency is required. In addition, this output impedance must be equal to the magnitude of load-line slope.

As can be seen from Fig. 7.5, the output voltage with ideal AVP can be expressed by Eq. (7.1), where R_{LL} is defined as load-line slope. It can be seen that VRs has constant output impedance which is equal to load-line slope R_{LL}, and CPU is modeled by an ideal current source.

$$V_{OUT} = V_{REF} - I_{OUT} \times R_{LL} \qquad (7.1)$$

7.2   Native Adaptive Voltage Positioning Design for Voltage Regulators

The AVP design of the adaptive on-time control circuit for VRs has a DC offset of the V_{OUT} voltage by Eq. (7.2), this Equation is not the same to the Eq. (7.1), so the AVP design of the adaptive on-time control circuit for VRs cannot achieve high accuracy the V_{OUT} voltage. Unfortunately, the VRs should be possessed a high accuracy the V_{OUT} voltage to avoid the V_{OUT} voltage may impact the load line specification.

$$V_{OUT} = \left(V_{VID} + \frac{\Delta V_{OUT}}{2} \right) - \frac{R_i}{A_V} \cdot \left(I_L - \frac{\Delta I_L}{2} \right) \qquad (7.2)$$

Native AVP design of the adaptive on-time control circuit for VRs can cancel the DC offset of the V_{OUT} voltage as shown in Fig. 7.6 [9, 30–41]. This control circuit not only has adaptive on-time control circuit, but also the ability to cancel the steady state error. The EA is an error amplifier and V_{COMP1} signal is the output of error amplifier. Two input signals of error amplifier are the feedback signal V_{FB} and

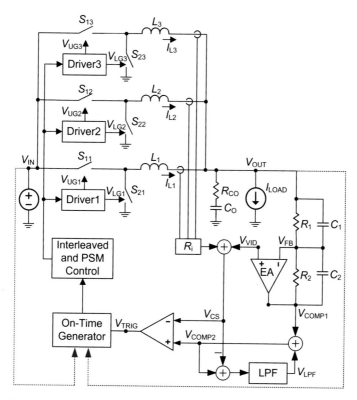

Fig. 7.6 Native AVP design of the adaptive on-time control circuit for VRs

the V_{VID} voltage, if the user wants to change the output voltage, and the user should be based on VR12 VID code table setting VID bits by I^2C interface [9, 30, 38, 41].

S_{11}, S_{12} and S_{13} are the upper-side switches, S_{21}, S_{22} and S_{23} are the lower-side switches, L_1, L_2, and L_3 are the output inductors, and R_{CO} is the ESR of the output capacitor C_O. The current source I_{LOAD} is the output load current. The feedback resistors, R_1 and R_2 can be designed by load line requirement, so the output voltage V_{OUT} is not equal to the V_{VID} voltage. The output voltage V_{OUT} depends on the output load current, if the system is operated from the light load to the heavy load and the output voltage V_{OUT} also should be reduced to meet the load line requirement. The trigger signal V_{TRIG} is the output of comparator to control on-time generator.

The steady state waveforms of the native AVP design are as shown in Fig. 7.7. A near-DC voltage is added with the output voltage V_{COMP1} of the error amplifier. The V_{COMP2} is lower than V_{COMP1} in steady state. The V_{OUT_NAVP} is without DC offset and it is also equal to expected value of the output voltage. The V_{OUT_OFS} voltage is a droop between the V_{OUT_NAVP} voltage and the V_{OUT_AVP} voltage. The V_{COMP_OFS} voltage is a droop between the V_{COMP1} voltage and V_{COMP2} voltage. The DC voltage is generated by a low pass filter subtraction of the V_{CS} and V_{COMP2}

Fig. 7.7 Steady state waveforms of the native AVP design

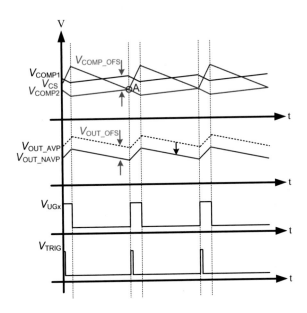

ripple voltages which are in front of the comparator. The native AVP design of the adaptive on-time control circuit for VRs can use a low pass filter subtraction to cancel the DC offset of the V_{OUT} voltage, so the native AVP design can achieve the high accuracy voltage V_{OUT} to meet load line specification. Moreover, the corner frequency of low pass filter should be designed much smaller than the switching frequency. Refer to the steady-state control signals at the "A" point shown in Fig. 7.7, and Eqs. (7.2)–(7.5) are obtained [9, 30, 38, 41].

$$V_{CS,Vally} = V_{COMP2,Peak} \qquad (7.2)$$

$$V_{VID} + R_i \cdot \left(I_L - \frac{\Delta I_L}{2} \right) = V_{COMP1} + V_{LPF} \qquad (7.3)$$

$$V_{VID} + R_i \cdot \left(I_L - \frac{\Delta I_L}{2} \right) = V_{VID} + A_V \cdot \left(V_{VID} - \left(V_{OUT} - \frac{\Delta V_{OUT}}{2} \right) \right) + V_{LPF} \qquad (7.4)$$

$$V_{LPF} = -A_V \frac{\Delta V_{OUT}}{2} - R_i \frac{\Delta I_L}{2} \qquad (7.5)$$

V_{LPF} is described in Eq. (7.5) and, by substituting (7.5) for (7.4). Equations (7.6)–(7.8) can be obtained.

$$V_{OUT} = V_{VID} - \frac{R_i}{A_V} I_L \qquad (7.6)$$

$$R_{DROOP} = \frac{R_i}{A_V} \qquad (7.7)$$

$$A_V = \frac{R_2}{R_1} \qquad (7.8)$$

A_V is the desired error amplifier gain. R_i is the internal current sense amplifier gain. R_{DROOP} is the current sense resistor. This control also implements the AVP function easily and solves a DC offset of the voltage V_{OUT}. R_{DROOP} should be designed to determine the load line. It is the equivalent load line resistance as well as the desired static output impedance.

An optimized compensation of a multiphase voltage regulator allows for the best possible load step response of its output. A type-II compensator with one pole and one zero is adequate for proper compensation. A prior design procedure shows how to decide the resistive feedback components of an error amplifier gain. C_1 and C_2 can be calculated for compensation by Eq. (7.9) [9, 30, 38, 41]. The target is to achieve constant resistive output impedance over the widest possible frequency range.

$$G_{VC}(s) = \frac{R_i}{R_{DROOP}} \frac{1 + s/(R_1 \times C_1)}{1 + s/(R_2 \times C_2)} \qquad (7.9)$$

7.3 On-Time Generator Circuit of the Constant Current Ripple On-Time Control Circuit for Voltage Regulators

On-time generator circuit of the constant current ripple on-time control circuit is as shown in Fig. 7.8, which samples the input voltage and the V_{VID} voltage to determine the width of the on-time. The TON is a pin of the IC through a resistor connected to the input voltage. It also samples the V_{VID} voltage by Eq. (7.10). The voltage drop between input voltage and the V_{VID} voltage depends on the inductor current ripple by Eq. (7.11). Therefore, even if the voltage drop between the input voltage and the V_{VID} voltage is changed, the inductor current ripple maintains a constant value. Hence, the constant current ripple on-time control circuit is suitable for different input and V_{VID} voltages to obtain the same output voltage ripple, such as CPU applications [9, 30–41].

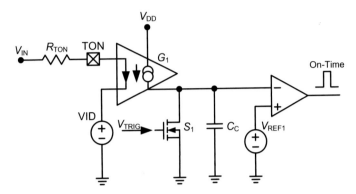

Fig. 7.8 On-time generator circuit of the constant current ripple on-time control circuit

$$T_{ON} = \frac{R_{TON}}{V_{IN} - VID} C_C \cdot V_{REF1} \tag{7.10}$$

$$T_{ON} = \frac{R_{TON}}{V_{IN} - V_{OUT}} \Delta I_L \tag{7.11}$$

When the resistor of R_{TON} is changed, the width of on-time is also changed. Switching frequency depends on the voltage drop between the input voltage and the V_{VID} voltage. Figure 7.9 shows the experimental results for the switching frequency versus the output load current of the constant current ripple on-time control circuit ($V_{IN} = 12$ V, $V_{VID} = 0.7$ V). Experimental results show that at an output load

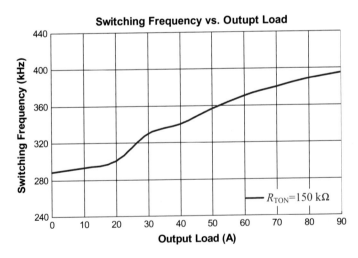

Fig. 7.9 Experimental results of the switching frequency versus the output load current for constant current ripple on-time control circuit

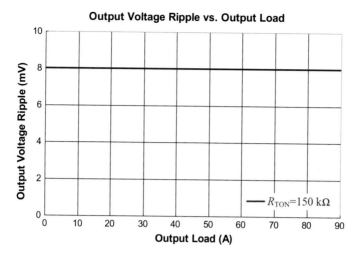

Fig. 7.10 Experimental results of the output voltage ripple versus the output load current for constant current ripple on-time control circuit

current of 0–90 A, when the resistor of R_{TON} is 150 kΩ, the switching frequency increases from 288 to 395 kHz.

Experimental results of the output voltage ripple versus the output load current for constant current ripple on-time control circuit are as shown in Fig. 7.10 (V_{IN} = 12 V, V_{VID} = 0.7 V). Experimental results show the operation at an output load current of 0–90 A. The inductor current ripple maintains a consistent value to obtain the same output voltage ripple, such as 8 mV.

Experimental results of the output voltage ripple for constant current ripple on-time control circuit are as shown in Fig. 7.11. The operation conditions are set at 0.7 V output voltage and 90 A output load current to measure the output voltage ripple for 12 V input voltage.

Fig. 7.11 Experimental results of the output voltage ripple for constant current ripple on-time control circuit

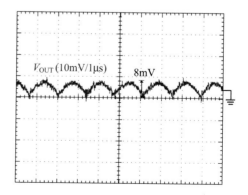

7.4   Comparison of Quick Dynamic Response and Conventional Quick Response of the On-Time Generator Circuit

CPU load transient may occur within 1 μs and have a large output load variation. Constant current ripple on-time control circuit also needed to add a non-linear open-loop quick response circuit to improve transient response. Further, most Constant current ripple on-time control circuit deliver energy from input terminal to CPU, try to solve this issue by setting up droop-voltage thresholds to trigger another open-loop regulation mechanism, such as triggering another on-time generator or increasing the duty of its on-time generator. However, this kind of design has two major drawbacks. First, the threshold is discrete, which means it may improve transient response over a specific threshold. Second, the threshold is fixed, which cannot meet the verity of loading conditions. Moreover, if the threshold can be set by external components, it suffers another drawback, extra pins, which increase cost and reduce the flexibility of board design.

Recently, two types of improved transient response have been used, namely, the internal fixed quick response by IC design, and the external setting by component or voltage source. The internal fixed quick response by IC design usually uses V_{FB} to determine and trigger the quick response function, which can provide longer on-time width to the driver circuit. The main advantage of this process is that no extra pin in the IC is needed to achieve this function. However, this generator circuit of the quick response function is not convenient to use in CPU applications because it cannot be designed to control the on-time width during different load conditions. On the other hand, the hysteresis of the comparator is hard to design, especially for very fast load transient like CPU applications. If the hysteresis of the comparator is too small, it can induce the system to trigger a longer on-time width erroneously, thereby causing a high ripple of V_{OUT}. Conversely, if the hysteresis of the comparator is too large, it can render the quick response function unavailable [9, 30–41].

External setting quick response of the on-time generator circuit for constant current ripple on-time control circuit is as shown in Fig. 7.12. The operational principle of the external setting quick response requires the addition of a voltage source V_{QRSET} to determine the width of QR-time. V_{QRSET} is connected to the "QRSET" pin. If the QR-time needs a longer on-time width, the user should design V_{QRSET} to be larger than or equal V_{REF1} [9, 30, 38, 41].

The QR trigger circuit samples the V_{OUT} signal and uses the low-pass filter to cause the V_{OUT} signal delay and a DC source of V_{QRTH}. At a steady state, the drop of V_{OUT} cannot trigger the QR-time. When this abrupt voltage drop is lower than the QR threshold level, the QR trigger circuit generates a low-level signal that turns off the switch of S2 as shown in Fig. 7.13. The frequency of the low-pass filter for R_{LOW} and C_{LOW} should be designed to be much smaller than the switching frequency which can avoid this system to have a failure in operation with QR-time. The V_{QRTH} signal can be designed in IC or set by the user for "QRSET" pin.

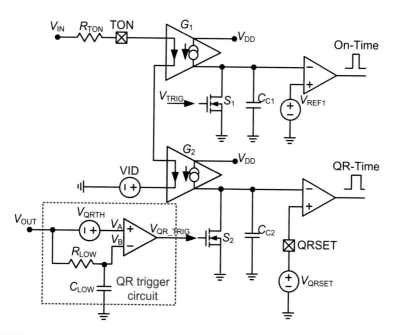

Fig. 7.12 External setting quick response of the on-time generator circuit for constant current ripple on-time control circuit

Fig. 7.13 Control signals for QR trigger circuit of external setting quick response

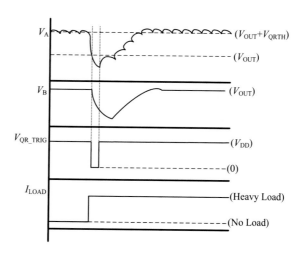

The main advantage of the external setting quick response can depend on the load conditions to design the V_{QRSET} value generating a suitable QR-time in this system. However, this method requires two extra IC pins to achieve the quick response function.

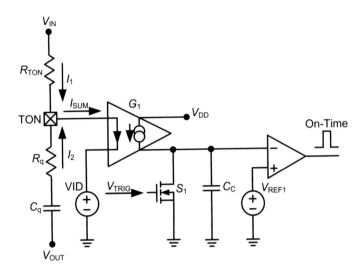

Fig. 7.14 Quick dynamic response of on-time generator circuit for constant current ripple on-time control circuit

Quick dynamic response of the constant current ripple on-time control circuit with native AVP design for voltage regulators is proposed as shown in Fig. 7.14. It is an external setting device that uses these components design, not a voltage source to set up. This quick dynamic response does not require any extra pin to achieve the quick response function of ICs. The system consists of resistor R_q and capacitor C_q in series that are connected between the V_{OUT} and TON pin. It operates under the principle that the high-pass filter filters high-frequency signals to pass from the V_{OUT} to the TON pin by Eq. (7.12). However, the frequency of the high-pass filter must be larger than or equal the switching frequency so the steady state operation of the system is not affected. Finally, the user may design the resistor R_q and the capacitor C_q based on the worst-case operation (maximum load step) [9, 30, 38, 41].

$$F_{RC} = \frac{1}{2\pi \cdot C_q \cdot R_q} \geq F_S \tag{7.12}$$

When light load quickly transforms quickly into a heavy load, V_{OUT} drops momentarily through the coupling by the capacitor C_q, which allows the resistor R_q to cause a voltage drop. The voltage drop of R_q forms a current of I_2, so the R_q needs to be designed first because the current of I_2 can change the width of on-time directly as shown in Fig. 7.15. However, a longer width of on-time makes the converter to deliver more energy from the input terminal to the output loading. If R_q has changed small, the width of on-time becomes longer. On the other hand, the capacitor of C_q should be designed with the frequency of the high-pass filter. The calculation of the Laplace transform formula is shown by Eq. (7.13).

Fig. 7.15 Control signals for quick dynamic response of on-time generator circuit for constant current ripple on-time control circuit

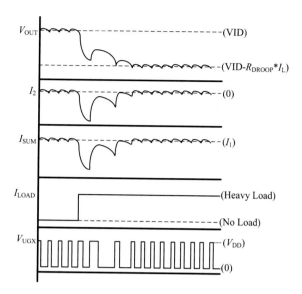

$$T_{ON}(s) = \frac{C_1}{G_1 \cdot \left((V_{IN} - VID)/R_{TON} + (s \cdot C_q \cdot V_{OUT} - VID)/(1 + s \cdot C_q \cdot R_q)\right)} V_{REFI}$$

(7.13)

In the multiphase operation, the quick dynamic response not only increases longer on-time width directly, but also controls the driver in each phase synchronously.

The advantages of the quick dynamic response are the follows:

1. It clearly generates a longer width of on-time that is proportional to the output voltage drop.
2. Adaption to the width of on-time depends on the load conditions of the system.
3. It only uses one generator circuit; thus, it is very convenient to design and apply.
4. It does not require an extra pin to achieve the quick dynamic response function.

7.5 Experimental Results

To understand the feasibility and the performance of the quick dynamic response of the constant current ripple on-time control circuit with native AVP design for voltage regulators, experimental results with multiphase operation are shown to prove that it is viable and useful. The operation conditions are as follows [9, 30, 38, 41]:

(1) Input voltage (V_{IN}): 12 V
(2) Output voltage (V_{VID}): 0.7 V

(3) Maximum output load current (I_{LOAD}): 35–75 A @ 75 A/μs
(4) Switching frequency (F_S): 288–395 kHz
(5) Upper-side MOSFET (S_{11}, S_{12}, S_{13}): IPS09N03LA * 3
(6) Lower-side MOSFET (S_{21}, S_{22}, S_{23}): IPS06N03LA * 6
(7) Compensation circuit: R_1 = 10 kΩ, R_2 = 47 kΩ, C_1 = 80 pF, C_2 = 100 pF
(8) Load line (R_{DROOP}): 2.25 mΩ
(9) Current Sense Amplifier Gain (R_i): 10
(10) Main inductor (L_1, L_2, L_3): 360 nH * 3
(11) Output capacitors (C_O): $C_{OS\text{-}CAP}$ = 560 μF/2.5 V (R_{CO}: 7 mΩ) * 3, $C_{MLCC\text{-}CAP}$ = 22 μF/6.3 V (R_{CO}: 3 mΩ) * 22
(12) Quick dynamic response (Fig. 7.8): G_1 = 147.54 m (A/A); R_{TON} = 150 kΩ; C_C = 3 pF; R_q = 330 Ω; C_q = 560 pF; V_{REF1} = 1.2 V
(13) External setting quick response (Fig. 7.6): R_{LOW} = 600 kΩ; C_{LOW} = 2 pF; G_1 = 147.54 m (A/A); R_{TON} = 150 kΩ; C_C = 3 pF; V_{REF1} = 1.2 V; V_{QRSET} = 1.2 V; V_{QRTH} = 35 mV

Figure 7.16 shows the chip layout of the control IC, in which the on-time generator and loop control are both occupied. The area is marked with a die size of 3040 μm × 2300 μm.

Experimental results at the droop with the quick dynamic response (Fig. 7.14) and without quick response (Fig. 7.8) are as shown in Fig. 7.17. The blue

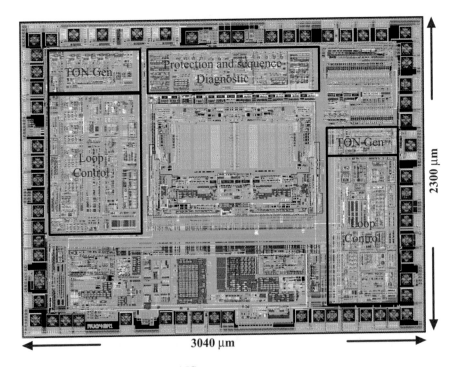

Fig. 7.16 Chip layout of the control IC

Fig. 7.17 Comparison of the experimental results at the droop with quick dynamic response and without quick response

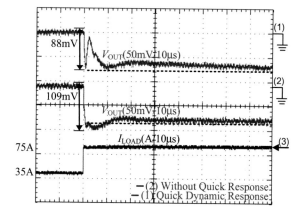

waveform has without quick response for the output voltage and the red waveform has the quick dynamic response for the output voltage. The output voltage signals are based on the same measured load current, such as the black waveform, which uses a voltage transient test (VTT) tool to achieve load transient [9, 30, 38, 41]. The VTT tool is created by Intel Corporation to simulate a CPU behavior, such as dynamic VID variation and dynamic load transient. The VTT tool is used to generate the maximum current slew rate of 300 A/μs for load transient.

V_{OUT} peak to peak value at the load transient with the quick dynamic response is 88 mV and V_{OUT} peak to peak value at the load transient without quick response is 109 mV. Hence, V_{OUT} with the quick dynamic response is not only lower by 21 mV than without quick response, but also has a lower voltage to meet the load line specification as shown by the black dot line. The load line specification has 90 mV from VR12 specification definition and it is also equal to the R_{DROOP} 2.25 mΩ to multiply the output loading step 40 A.

Experimental results of the output voltage and control signals at the droop without quick response are as shown in Fig. 7.18. The blue waveform is without

Fig. 7.18 Experimental results of the output voltage and control signals at the droop without quick response

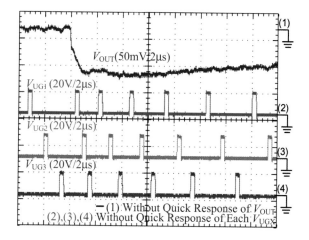

quick response for the output voltage, the green waveform is without quick response for the V_{UG1} signal, the orange waveform is without quick response for the V_{UG2} signal, the pink waveform is without quick response for the V_{UG3} signal. These control signals have the same width of on-time and a sequencing control. On the other hand, these control signals cannot drive each phase synchronously to deliver more energy from the input terminal to the output loading.

Experimental results of the output voltage and control signals at the droop with quick dynamic response are as shown in Fig. 7.19. The red waveform is that with quick dynamic response for the output voltage signal, the green waveform is that with quick dynamic response for the V_{UG1} signal, the orange waveform is that with quick dynamic response for the V_{UG2} signal, the pink waveform is that with quick dynamic response for the V_{UG3} signal. These control signals with quick dynamic response are obviously longer than those without quick response during the output loading, from light load to heavy load. On the other hand, these control signals can drive all phases synchronously to deliver more energy from the input terminal to the output loading, the output voltage has increased after quick dynamic response triggered.

Simulation results of the output voltage and control signals at the droop with quick dynamic response are as shown in Fig. 7.20. The red waveform is that with quick dynamic response for the output voltage signal, the green waveform is that with quick dynamic response for the V_{UG1} signal, the orange waveform is that with quick dynamic response for the V_{UG2} signal, the pink waveform is that with quick dynamic response for the V_{UG3} signal. These control signals with quick dynamic response are obviously longer than those without quick response during the output loading, from light load to heavy load. On the other hand, these control signals can drive all phases synchronously to deliver more energy from the input terminal to the output loading to prevent the output voltage from dropping significantly. Thus, SIMPLIS simulation and experimental results are matched.

Fig. 7.19 Experimental results of the output voltage and control signals at the droop with quick dynamic response

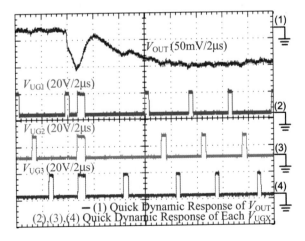

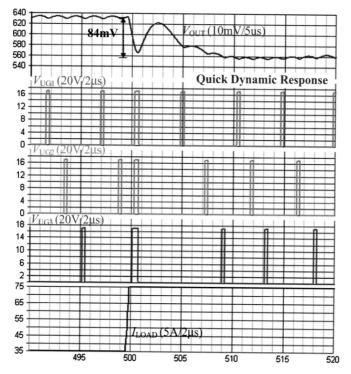

Fig. 7.20 Simulation results of the output voltage and control signals at the droop with quick dynamic response

To further understand the advantage and the superiority of the quick dynamic response, the plans to base identical operation conditions by comparing the external setting quick response.

Experimental results at the droop with quick dynamic response (Fig. 7.14) and external setting quick response (Fig. 7.12) are as shown in Fig. 7.21. The blue waveform is that with external setting quick response for the output voltage, the red waveform is that with quick dynamic response for the output voltage. The output voltage signals are based on the same measured load current, such as the black waveform used as VTT tool to achieve load transient. V_{OUT} peak to peak value at the droop with quick dynamic response is 88 mV and V_{OUT} peak to peak value at the droop with external setting quick response is 96 mV. Thus, V_{OUT} with quick dynamic response and external setting quick response are close. Quick dynamic response and external setting quick response are useful to prevent the V_{OUT} from dropping significantly, but the quick dynamic response does not require an extra pin to achieve the quick response function.

Fig. 7.21 Comparison of the experimental results at the droop with quick dynamic response and external setting quick response

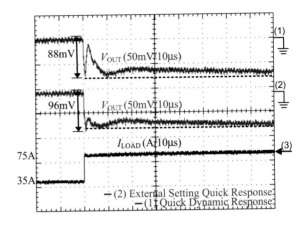

7.6   Summary

This chapter proposes to achieve quick dynamic response of the constant current ripple on-time control circuit with native AVP design for voltage regulators. The concept uses the capacitor and resistor in series to filter V_{OUT} at load transient to change the width of on-time dynamically to prevent the V_{OUT} from dropping significantly.

Both experimental and simulation results confirm that the proposed the quick dynamic response of the constant current ripple on-time control circuit with native AVP design for voltage regulators can significantly improve transient response. Moreover, the proposed quick dynamic response has very simple structure and design with native AVP design for voltage regulators.

References

1. E. Stanford, Power technology roadmap for microprocessor voltage regulators, in *Presentation at PSMA*, Feb 2003
2. 2001 technology roadmap for semiconductors, available: http://www.intel.com/research/silicon/AlanAllanIEEEComputer0102.pdf
3. K. Yao, Y. Ren, J. Sun, K. Lee, M. Xu, J. Zhou, F.C. Lee, Adaptive voltage position design for voltage regulators, in *Proceedings of IEEE Applied Power Electronics Conference and Exposition Conference* (2004), pp. 272–278
4. H. Mao, L. Yao, C. Wang, I. Batarseh, Analysis of Inductor current sharing in nonisolated and isolated Multiphase DC–DC converters. IEEE Trans. Ind. Electron. **54**(6), 3379–3388 (2007)
5. P.-W. Lee, Y.-S. Lee, D.K.W. Cheng, X.-C. Liu, Steady-state analysis of an interleaved boost converter with coupled inductors. IEEE Trans. Ind. Electron. **47**(4), 787–795 (2000)
6. H.N. Nagaraja, D. Kastha, A. Patra, Design principles of a symmetrically coupled Inductor structure for Multiphase synchronous Buck converters. IEEE Trans. Ind. Electron. **58**(3), 988–997 (2011)
7. L.-P. Wong, D.K.-W. Cheng, M.H.L. Chow, Y.-S. Lee, Interleaved three-phase forward converter using integrated transformer. IEEE Trans. Ind. Electron. **52**(5), 1246–1260 (2005)

8. J. Abu-Qahouq, H. Mao, I. Batarseh, Multiphase voltage-mode hysteretic controlled DC–DC converter with novel current sharing. IEEE Trans. Power Electron. **19**(6), 1397–1407 (2004)
9. W.W. Chen, J.F. Chen, T.J. Liang, S.F. Hsiao, J.R. Huang, W.Y. Ting, Improved transient response using HFFC circuit of the CCRCOT with native AVP design for voltage regulators. IET Power Electron. **6**, 1948–1955 (2013)
10. C. Ni, T. Tetsuo, Adaptive constant on-time (D-CAP™) control study in notebook applications, in *Texas Instruments, Application report SLVA281B*, July 2007
11. R. Redl, J. Sun, Ripple-based control of switching regulators an overview. IEEE Trans. Power Electron. **24**, 2669–2680 (2009)
12. J. Li, Current-mode control: modeling and its digital application, Ph.D. thesis, Virginia Polytechnic Institute and State University, 2009
13. J. Sun, Characterization and performance comparison of ripple-based control for voltage regulator modules. IEEE Trans. Power Electron. **21**, 346–353 (2006)
14. W. Huang, A new control for multi-Phase buck converter with fast transient response, in *Proceedings of IEEE Applied Power Electronics Conference and Exposition Conference* (2001), pp. 273–279
15. J. Li, F.C. Lee, Modeling of V² current-mode control, in *Proceedings of IEEE Applied Power Electronics Conference and Exposition Conference* (2009), pp. 298–304
16. S.J. Wang, Y.H. Lee, Y.C. Lai, K.H. Chen, Quadratic differential and integration technique in V2 control buck converter with small ESR capacitor, in *Proceedings of IEEE Custom Integrated Circuits Conference* (2009), pp. 211–214
17. R. Redl, G. Reizik, Switched noise filter for the buck converter using the output ripple as the PWM ramp, in *Proceedings of IEEE Application Power Electronics Conference* (2005), pp. 918–924
18. J. Li, F.C. Lee, New Modeling Approach and Equivalent Circuit Representation for Current-Mode Control. IEEE Trans. Power Electron. **25**, 1218–1230 (2010)
19. Y.H. Lee, S.J. Wang, K.H. Chen, Quadratic differential and integration technique in V² control buck converter with small ESR capacitor. IEEE Trans. Power Electron. **25**, 829–838 (2010)
20. M.Y. Yen, P. Mok, A constant frequency output-ripple voltage-based buck converter without using large ESR capacitor. IEEE Trans. Circuits Syst. I **55**, 748–752 (2008)
21. K.Y. Cheng, F. Yu, P. Mattavelli, F.C. Lee, Characterization and performance comparison of digital V²-type constant on-time control for buck converters, in *IEEE Control and Modeling for Power Electronics Conference*, June 2010, pp. 1–6
22. A.V. Petershevs, S.R. Sanders, Digital multimode buck converter control with loss-minimizing synchronous rectifier adaptation. IEEE Trans. Power Electron. **21**(6), 1588–1599 (2006)
23. J. Wang, B. Bao, G. Zhou, W. Hu, Dynamical effects of equivalent series resistance of output capacitor in constant on-time controlled Buck converter. IEEE Trans. Ind. Electron. **58**(12), 5406–5410 (2011)
24. S. Angkititrakul, H. Hu, Design and analysis of buck converter with pulse-skipping modulation, in *Proceedings of IEEE Power Electronics Specialists Conference* (2008), pp. 1151–1156
25. X. Zhou, M. Donati, L. Amoroso, F.C. Lee, Improved light-load efficiency for synchronous rectifier voltage regulator module. IEEE Trans. Power Electron. **15**(5), 826–834 (2000)
26. C.L. Chen, W.L. Hsieh, W.J. Lai, K.H. Chen, C.S. Wang, A new PWM/PFM control technique for improving efficiency over wide load range, in *Proceedings of IEEE International Conference on Electronics, Circuits and Systems*, Aug 2008, pp. 962–965
27. S. Kapat, S. Banerjee, A. Patra, Discontinuous map analysis of a DC-DC converter governed by pulse skipping modulation. IEEE Trans. Circuits Syst. I Part I **57**(7), 1793–1801 (2010)
28. C.A. Yeh, Y.S. Lai, Digital pulsewidth modulation technique for a synchronous Buck DC/DC converter to reduce switching frequency. IEEE Trans. Ind. Electron. **59**(1), 550–561 (2012)
29. J. Wang, J. Xu, B. Bao, Analysis of pulse bursting phenomenon in Constant-On-Time-Controlled Buck converter. IEEE Trans. Ind. Electron. **58**(12), 5406–5410 (2011)

30. J.R. Huang, C.H. Wang, C.J. Lee, K.L. Tseng, D. Chen, Native AVP control method for constant output impedance of DC power converters, in *Proceedings of IEEE Power Electronics Specialists Conference* (2007), pp. 2023–2028
31. A. Waizman, C.Y. Chung, Resonant free power network design using extended adaptive voltage positioning (EAVP) methodology. IEEE Trans. Adv. Packag. **24**, 236–244 (2001)
32. P.L. Wong, Performance improvements of multi-channel interleaving voltage regulator modules with integrated coupling inductors, Dissertation of Virginia Polytechnic Institute and State University, Mar 2001
33. M. Lee, D. Chen, K. Huang, E. Tseng, B. Tai, Compensator Design for Adaptive Voltage Position (AVP) for Multiphase VRMs, in *Proceedings of IEEE Power Electronics Specialists Conference* (2006)
34. K. Yao, Y. Meng, P. Xu, F.C. Lee, Design considerations for VRM transient response based on the output impedance, in *Proceedings of IEEE Applied Power Electronics Conference and Exposition Conference* (2002), pp. 14–20
35. R. Redl, N.O. Sokal, Near-optimum dynamic regulation of DC-DC converters using feed-forward of output current and input voltage with current-mode control. IEEE Trans. Power Electron. **1**, 181–192 (1986)
36. S.K. Mishra, Design-oriented analysis of modern active droop controlled power supplies. IEEE Trans. Ind. Electron. **56**(9), 3704–3708 (2009)
37. J.A.A. Qahouq, V. Arikatla, Power converter with digital sensorless adaptive voltage positioning control scheme. IEEE Trans. Ind. Electron. **58**(9), 4105–4116 (2010)
38. Richtek Technology Corporation, Dual output 3-Phase + 2-Phase PWM controller for CPU and GPU Core power supply, RT8885A Datasheet, 2012
39. K.P. Liu, K.C. Wang, L.P. Tai, C.S. Cheng, Control circuit and method for a constant on-time PWM switching converter, U. S. Patent 7 834 606 B2, 2 Aug 2007
40. K Huang, Spring modulation with fast load-transient response for a voltage regulator, U. S. Patent 7 247 182 B2, 26 Aug 2005
41. Richtek Technology Corporation, Multi-Phase PWM controller for CPU core power supply, RT8859M Datasheet, 2014